Understanding Aquatic Ecology

Understanding Aquatic Ecology

Vincent Jennings

Larsen & Keller
www.larsen-keller.com

Understanding Aquatic Ecology
Vincent Jennings
ISBN: 978-1-64172-668-9 (Hardback)

Larsen & Keller
Published by Larsen and Keller Education,
5 Penn Plaza,
19th Floor,
New York, NY 10001, USA

Cataloging-in-Publication Data

Understanding aquatic ecology / Vincent Jennings.
p. cm.
Includes bibliographical references and index.
ISBN 978-1-64172-668-9
1. Aquatic ecology. 2. Ecology. 3. Aquatic biology. 4. Aquatic habitats. I. Jennings, Vincent.
QH541.5.W3 U53 2022
577.6--dc23

For more information regarding Larsen and Keller Education and its products, please visit the publisher's website www.larsen-keller.com

Table of Contents

Preface

The branch of ecology which studies the interactions between various organisms which dwell in water bodies, and their surroundings, is known as aquatic ecology. The different aquatic environments which fall under the domain of aquatic ecology are estuaries, lakes, ponds, oceans and rivers. An ecosystem which exists in a body of water is known as an aquatic ecosystem. There are primarily two types of aquatic ecosystems, marine and freshwater ecosystems. Marine ecosystems are further divided into various zones like oceanic zone, benthic zone and intertidal zone. Freshwater ecosystems are broadly divided into three types - lentic, lotic and wetlands. Lentic ecosystems are made up of slow moving water. Lotic ecosystems consist of faster moving water, such as streams and rivers. Wetlands are ecosystems which are flooded by water, either seasonally or permanently. This book provides comprehensive insights into the field of aquatic ecology. The topics included herein on this subject are of utmost significance and bound to provide incredible insights to readers. Those in search of information to further their knowledge will be greatly assisted by this book.

To facilitate a deeper understanding of the contents of this book a short introduction of every chapter is written below:

Chapter 1- The study of the relationships between individual organisms within a species and between different species in aquatic environments is termed as aquatic ecology. It is also involved in the study of the interactions between organisms and their physical and chemical environments. This is an introductory chapter which will introduce briefly all the significant aspects of ecology and aquatic ecology.

Chapter 2- The study of the different organisms and the relationships among them as well as with the environment in aquatic ecosystems which have a high salt content is termed as marine ecology. Some of the marine ecosystems which are studied within this field are salt marshes, mangroves, estuaries and lagoons. This chapter discusses in detail these marine ecosystems as well as the threats to them.

Chapter 3- The branch of ecology which deals with the study of flowing waters as well as the various abiotic and biotic interactions among animals, plants and microorganisms within rivers is termed as river ecology. The diverse aspects of river ecology such as the restorations of rivers have been thoroughly discussed in this chapter.

Chapter 4- The branch of ecology which focuses on the study of still water ecosystems is termed as lake ecology. Some of the aspects studied in relation to this field are watershed ecology and the different lake zones. This chapter closely examines these key concepts of lake ecology as well as the benefits of lakes to provide an extensive understanding of the subject.

Chapter 5- Wetlands are ecosystems which are flooded by water and which contains aquatic plants adapted to the unique hydric soil. The topics elaborated in this chapter will help in gaining a better perspective about the conservation of wetlands as well as the different plants, birds and other organisms found in wetlands.

I would like to share the credit of this book with my editorial team who worked tirelessly on this book. I owe the completion of this book to the never-ending support of my family, who supported me throughout the project.

Vincent Jennings

1

Introduction to Aquatic Ecology

The study of the relationships between individual organisms within a species and between different species in aquatic environments is termed as aquatic ecology. It is also involved in the study of the interactions between organisms and their physical and chemical environments. This is an introductory chapter which will introduce briefly all the significant aspects of ecology and aquatic ecology.

Ecology

Ecology is the branch of biology that studies how organisms interact with their environment and other organisms. Every organism experiences complex relationships with other organisms of its species, and organisms of different species. These complex interactions lead to different selective pressures on organisms. The pressures together lead to natural selection which causes populations of species to evolve. Ecology is the study of these forces, what produces them, and the complex relationships between organisms and each other, and organisms and their non-living environment.

The field of ecology has a huge variety of sub-disciples. Although the types of ecology below are divided by the level of organization being viewed, some ecologist specializes in specific aspects of each field. For example, cognitive ecology is a branch of organismal ecology that studies the mental processes of animals. Other ecologists study only the interactions between humans and the rest of the biome, a field known as human ecology or environmental science. Still other ecologists focus on the interactions between organisms and the abiotic factors that affect their evolution, such as nutrients and toxins in the environment. Ecology is a vast field, the following are only a sampling of the variety of things scientist study in ecology.

Three Basic Types of Ecology

All ecology breaks down and falls under three basic types. Under each of the basic types you can break down the ecologies even further, but these are the ones that will give you an understanding of what is an ecology.

1. Conservation Ecology – This refers to the ecology of the natural world and how it exists without the presence or interference of man, and how it exists with the presence and interference of man. It is common to find different areas in the wilderness or portions of public parks set aside with entrance restricted or forbidden because they are conservation areas. Conservation areas are set aside to protect conservation ecologies.

2. Urban Ecology – As you can guess by the name this one is focused on the balance of life within urban settings. It includes both conservation and human ecology, but the nature of an urban ecology is so different in how phases of ecological succession may develop that it is studied as

its own entity. Urban ecology looks at the impact of human life on an area, how cities and urban areas manage resources and what the cycle of growth and decay within these unique environments are.

3. Human Ecology – Human ecology is multi-faceted. It focuses greatly on patterns of population and mortality, consumption of resources, conservation efforts and how humans affect plant, animal and other human life. Within human ecology you will also find people studying the impact of the human race on the atmosphere, space and the ocean.

Three Phases of Ecological Succession

Each of these ecology types go through three distinct phases of ecological succession. Succession is the term used to define how the development in one phase of the ecology is then used to allow the next phase to occur. While it may seem like this is a very ordered process, if the ecology is disrupted it may go back or forward a phase. To better explain succession we will use the example of what happens after a forest fire has destroyed life on the side of a mountain.

Primary: This is the beginning stage of an ecology. That side of the mountain may begin to be repopulated with seeds that have blown in on the wind, been planted by humans in a conservation effort, and the area may gain animal life from different types moving in to hunt or live for protection. The key to understanding primary ecology is that is begins with the minimal life that the area can support, which is then followed by the next life in the food or resource chain.

Secondary: During the secondary phase of the ecology life has begun again in the forest area. Grass and flowers may grow, small birds and mammals have returned and there have been two or three cycles of life to seed to decay to birth that have occurred. This then brings in more predators in the cycle to complete the cycle of life.

Climax: The climax stage of ecological succession sees a forest thriving on the mountainside again with all the plant and animal life you would expect. There is little evidence of the fire as the forest has aged enough to still be young, but to be well established. It is at this point that the ecology begins to slowly decline as it begins to overuse resources and conservation is needed.

Disruptions

Problems happen when an ecology is disrupted. This can speed an ecology through its successive phases and cause it to die out quickly. Remember that no ecology exists by itself, but all ecologies exist within balance to each other. A small conservation ecology in the rainforest may not seem like a big deal, but it plays a part in managing air, climate and other resources needed by the human ecology a continent away.

Natural Disruptions

Natural disruptions can come in the form of extreme weather such as prolonged droughts, heavy rain or snowfall, storms, hurricanes, tornado, monsoons and more. These also can involve seismic disruptions such as earthquakes and volcanic eruptions. These dramatic events can damage or change the balance within an ecology.

Man-made Disruptions

Man-made disruptions can come in all shapes and sizes. There are the obvious ones such as clearing land, but less obvious ones when you are looking at an ecology in its secondary stage of succession that includes mankind. An increased drain on natural resources, such as water or fossil fuels, can set up a disruption. All forms of mining, including wind farms, disrupt ecologies as well.

Protection of Ecology

There are many different ways that an ecology can be protected. Most of them are in place or in the process of being put in place now as the world begins to understand the importance of managing ecologies better. The three main ways that an ecology is protected are:

Conservation: As mentioned before, conservation is when an ecology is protected by harm by being set aside so it cannot be interfered with. In the USA, many of the major state and national parks are conservation areas. You can only visit certain places in the park while other areas are off limits to protect the ecology there.

Regulation: Regulations such as the Clean Air Act and other laws try to set limits on the types of man-made disruptions caused by expansion, exploration or industrialization that can occur. The goal is to limit the known impact on the natural ecology to preserve and maintain the resources that mankind depends on.

Replacement: This type of replacement is twofold. In areas where regulation has permitted conservation to be removed, companies may pay to have a new area seeded or protected to try and restore balance. Another aspect of replacement is the search for man-made substances that can replace natural resources. One example of this is the search for a viable alternative energy to fossil fuel.

Aquatic Ecology

An aquatic ecosystem includes a group of interacting organisms which are dependent on one another and their water environment for nutrients and shelter. Examples of aquatic ecosystem include oceans, lakes and rivers.

An aquatic ecosystem includes freshwater habitats like lakes, ponds, rivers, oceans and streams, wetlands, swamp, etc. and marine habitats include oceans, intertidal zone, reefs, seabed and so on. The aquatic ecosystem is the habitat for water-dependent living species including animals, plants, and microbes.

Types of Aquatic Ecosystem

Freshwater Aquatic Ecosystem

They cover only a small portion of earth nearly 0.8 per cent. Freshwater involves lakes, ponds, rivers and streams, wetlands, swamp, bog and temporary pools. Freshwater habitats are classified into lotic and lentic habitats. Water bodies such as lakes, ponds, pools, bogs, and other reservoirs are standing water and known as lentic habitats. Whereas lotic habitats represent flowing water bodies such as rivers, streams.

Lotic Ecosystems: They mainly refer to the rapidly flowing waters that move in a unidirectional way including the rivers and streams. These environments harbor numerous species of insects such as beetles, mayflies, stoneflies and several species of fishes including trout, eel, minnow, etc. Apart from these aquatic species, these ecosystems also include various mammals such as beavers, river dolphins and otters.

Lentic Ecosystems: They include all standing water habitats. Lakes and ponds are the main examples of Lentic Ecosystem. The word lentic mainly refers to stationary or relatively still water. These ecosystems are home to algae, crabs, shrimps, amphibians such as frogs and salamanders, for both rooted and floating-leaved plants and reptiles including alligators and other water snakes are also found here.

Wetlands: Wetlands are marshy areas and are sometimes covered in water which has a wide diversity of plants and animals. Swamps, marshes, bogs, black spruce and water lilies are some examples in the plant species found in the wetlands. The animal life of this ecosystem consists of dragonflies and damselflies, birds such as Green Heron and fishes such as Northern Pike.

Marine Aquatic Ecosystem

Marine ecosystem covers the largest surface area of the earth. Two third of earth is covered by water and they constitute of oceans, seas, intertidal zone, reefs, seabed, estuaries, hydrothermal vents and rock pools. Each life form is unique and native to its habitat. This is because they have adaptations according to their habitat. In the case of aquatic animals, they can't survive outside of water. Exceptional cases are still there which shows another example of adaptations (e.g. mudskippers). The marine ecosystem is more concentrated with salts which make it difficult for freshwater organisms to live in. Also, marine animals cannot survive in freshwater. Their body is adapted to live in saltwater; if they are placed in less salty water, their body will swell (osmosis).

Ocean Ecosystems

Our planet earth is gifted with the five major oceans, namely Pacific, Indian, Arctic, and the Atlantic Ocean. Among all these five oceans, the Pacific and the Atlantic are the largest and deepest ocean. These oceans serve as a home to more than five lakh aquatic species. Few creatures of these ecosystems include shellfish, shark, tube worms, crab small and large ocean fishes, turtles, crustaceans, blue whale, reptiles, marine mammals, seabirds, plankton, corals and other ocean plants.

Coastal Systems

They are the open systems of land and water which are joined together to form the coastal ecosystems. The coastal ecosystems have a different structure, and diversity. A wide variety of species of aquatic plants and algae are found at the bottom of the coastal ecosystem. The fauna is diverse and it mainly consists of crabs, fish, insects, lobsters snails, shrimp, etc.

Plants and animals in an aquatic ecosystem show a wide variety of adaptations which may involve life cycle, physiological, structural and behavioral adaptations. Majority of aquatic animals are streamlined which helps them to reduce friction and thus save energy. Fins and gills are the locomotors and respiratory organs respectively. Special features in freshwater organisms help them to drain excess water from the body. Aquatic plants have different types of roots which help them to survive in water. Some may have submerged roots; some have emergent roots or maybe floating plants like water hyacinths.

Limnology

Limnology is the study of inland waters - lakes (both freshwater and saline), reservoirs, rivers, streams, wetlands, and groundwater - as ecological systems interacting with their drainage basins and the atmosphere. The limnological discipline integrates the functional relationships of growth, adaptation, nutrient cycles, and biological productivity with species composition, and describes and evaluates how physical, chemical, and biological environments regulate these relationships. Stated simply, limnology is the study of the structural and functional interrelationships of organisms of inland waters as their dynamic physical, chemical, and biotic environments affect them.

Freshwater ecology is the study of the structure, function, and change of organisms in fresh waters as affected by their dynamic physical, chemical, and biotic environments.

Freshwater biology is the study of the biological characteristics and interactions of organisms of fresh waters. This study is largely restricted to the organisms themselves, such as their biology, life histories, populations, or communities.

Limnology encompasses an integration of physical, chemical, and biological components of inland aquatic ecosystems with the drainage basin, movements of water through the drainage basin, and biogeochemical changes that occur en route, and within standing (lentic) waters and exchanges with the atmosphere. The lake ecosystem is intimately coupled with its drainage area and atmosphere, and with its running (lotic) waters and ground waters that flow, and metabolize en route, components of the land being transported to the lake.

Understanding of the causal mechanisms operating in and controlling our natural world is a primary objective of limnology because of the premier importance of fresh water for the wellbeing of humankind. The greater our understanding, the higher the probability to predict accurately patterns of events within aquatic ecosystems in response to human manipulations and disturbances. A combination of analytical techniques is used to acquire that understanding:

1. Descriptive observations of patterns of biological processes and communities in relation to dynamic patterns of environmental properties. Such descriptive empirical analyses allow the generation of hypotheses, that is, conceptual predictive "models" of relationships among observed patterns.

2. Experimental examination and evaluation of quantitative responses to selected disturbances imposed on the system. By imposing quantitatively known disturbances on specific parts of the community or ecosystem, much insight can be gained on controlling factors governing their operation. In some cases, entire lakes or streams are experimentally manipulated.

3. Application of quantitative predictive models based on experimentally established, not random, governing variables. Models allow expansion of experimentally understood quantitative relationships, that is, hypothetical data can be inserted allowing a theoretical estimate of system responses to these variables.

References

- Ecology: biologydictionary.net, Retrieved 19 May, 2019
- What-is-ecology: conserve-energy-future.com, Retrieved 28 April, 2019
- Aquatic-ecosystem, biology: byjus.com, Retrieved 21 July, 2019
- Limnology, page: aslo.org, Retrieved 7 May, 2019

2

Marine Ecology

The study of the different organisms and the relationships among them as well as with the environment in aquatic ecosystems which have a high salt content is termed as marine ecology. Some of the marine ecosystems which are studied within this field are salt marshes, mangroves, estuaries and lagoons. This chapter discusses in detail these marine ecosystems as well as the threats to them.

Marine Ecology is the scientific study of marine-life habitat, populations, and interactions among organisms and the surrounding environment including their abiotic (non-living physical and chemical factors that affect the ability of organisms to survive and reproduce) and biotic factors (living things or the materials that directly or indirectly affect an organism in its environment).

Marine ecology is a subset of the study of marine biology and includes observations at the biochemical, cellular, individual, and community levels as well as the study of marine ecosystems and the biosphere.

The study of marine ecology also includes the influence of geology, geography, meteorology, pedology, chemistry, and physics on marine environments. The impact of human activity such as medical research, development, agriculture, fisheries, and forestry is also studied under marine ecology. In some ways, marine ecology is more complex than the relatively straightforward study of a particular organism or environment because of the numerous interconnections, symbiotic relationships, and influence of many factors on a particular environment.

To understand the difference between marine biology and marine ecology, it may be useful to look at a community of organisms. A marine biologist may focus on behavioral relationships between the organisms in one particular species while someone studying ecology would study how the behavior of one organism influences another. An ecologist would also look at abiotic factors and how they influence that organism. A scientist studying community ecology might study a group of organisms to see how they influence other species and abiotic factors.

Marine Life

The phrase 'marine life' refers to organisms that live in salt water. These can include a diverse array of plants, animals and microbes (tiny organisms) such as bacteria and archaea.

From the perspective of a land animal like us, the ocean can be a harsh environment. However, marine life are adapted to live in the ocean. Characteristics that help marine life thrive in a saltwater environment include the ability to regulate their salt intake or deal with large quantities of salt water, adaptations to obtain oxygen (e.g. a fish's gills), being able to withstand high water pressures, living in a place where they can get enough light, or being able to adjust to a lack of light. Animals and plants that live on the edge of the ocean, such as tide pool animals and plants, also need to deal with extremes in water temperature, sunlight, wind and waves.

Microorganisms

Microbial Mats	

Microbial mats are the earliest form of life on Earth for which there is good fossil evidence. The image shows a cyanobacterial-algal mat.	Stromatolites are formed from microbial mats as microbes slowly move upwards to avoid being smothered by sediment.

Microorganisms make up about 70% of the marine biomass. A microorganism, or microbe, is a microscopic organism too small to be recognised with the naked eye. It can be single-celled or multicellular. Microorganisms are diverse and include all bacteria and archaea, most protozoa such as algae, fungi and certain microscopic animals such as rotifers.

Many macroscopic animals and plants have microscopic juvenile stages. Some microbiologists also classify viruses (and viroids) as microorganisms, but others consider these as nonliving.

Microorganisms are crucial to nutrient recycling in ecosystems as they act as decomposers. Some microorganisms are pathogenic, causing disease and even death in plants and animals. As inhabitants of the largest environment on Earth, microbial marine systems drive changes in every global system. Microbes are responsible for virtually all the photosynthesis that occurs in the ocean, as well as the cycling of carbon, nitrogen, phosphorus and other nutrients and trace elements.

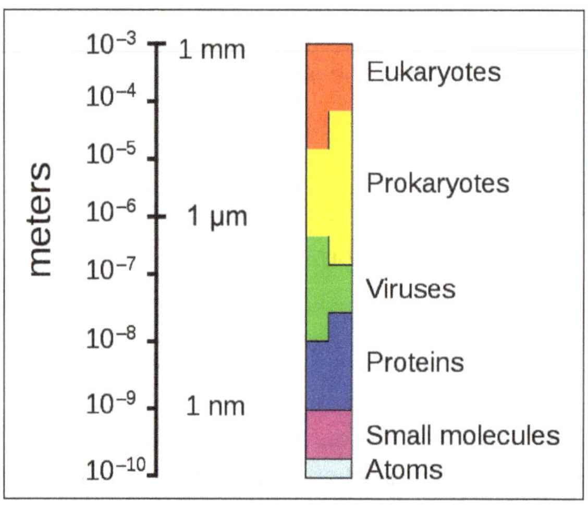

The range of sizes shown by prokaryotes (bacteria and archaea) and viruses relative to those of other organisms and biomolecules.

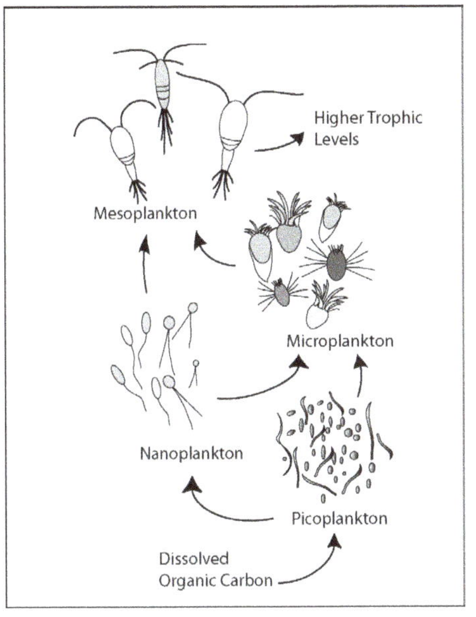

Marine microbial loop.

Microscopic life undersea is diverse and still poorly understood, such as for the role of viruses in marine ecosystems. Most marine viruses are bacteriophages, which are harmless to plants and animals, but are essential to the regulation of saltwater and freshwater ecosystems. They infect and destroy bacteria in aquatic microbial communities, and are the most important mechanism of recycling carbon in the marine environment. The organic molecules released from the dead bacterial cells stimulate fresh bacterial and algal growth. Viral activity may also contribute to the biological pump, the process whereby carbon is sequestered in the deep ocean.

Sea spray containing marine microorganisms can be swept high into the atmosphere and may travel the globe before falling back to earth.

A stream of airborne microorganisms circles the planet above weather systems but below commercial air lanes. Some peripatetic microorganisms are swept up from terrestrial dust storms, but most originate from marine microorganisms in sea spray. In 2018, scientists reported that hundreds of millions of viruses and tens of millions of bacteria are deposited daily on every square meter around the planet.

Microscopic organisms live throughout the biosphere. The mass of prokaryote microorganisms — which includes bacteria and archaea, but not the nucleated eukaryote microorganisms — may be as much as 0.8 trillion tons of carbon (of the total biosphere mass, estimated at between 1 and 4 trillion tons). Single-celled barophilic marine microbes have been found at a depth of 10,900 m (35,800 ft) in the Mariana Trench, the deepest spot in the Earth's oceans. Microorganisms live inside rocks 580 m (1,900 ft) below the sea floor under 2,590 m (8,500 ft) of ocean off the coast of the northwestern United States, as well as 2,400 m (7,900 ft; 1.5 mi) beneath the seabed off Japan. The greatest known temperature at which microbial life can exist is 122 °C (252 °F) (*Methanopyrus kandleri*). In 2014, scientists confirmed the existence of microorganisms living 800 m (2,600 ft) below the ice of Antarctica. According to one researcher, "You can find microbes everywhere — they're extremely adaptable to conditions, and survive wherever they are."

Under a magnifier, a splash of seawater teems with life.

Marine Viruses

Viruses are small infectious agents that do not have their own metabolism and can replicate only inside the living cells of other organisms. Viruses can infect all types of life forms, from animals and plants to microorganisms, including bacteria and archaea. The linear size of the average virus is about one one-hundredth that of the average bacterium. Most viruses cannot be seen with an optical microscope so electron microscopes are used instead.

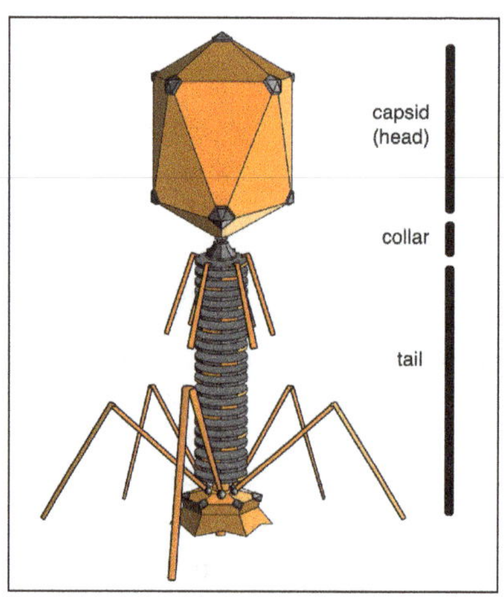

Diagram of a typical tailed phage.

Viruses are found wherever there is life and have probably existed since living cells first evolved. The origin of viruses is unclear because they do not form fossils, so molecular techniques have been used to compare the DNA or RNA of viruses and are a useful means of investigating how they arise.

Viruses are now recognised as ancient and as having origins that pre-date the divergence of life into the three domains. But the origins of viruses in the of life are unclear: some may have evolved from plasmids—pieces of DNA that can move between cells—while others may have evolved from bacteria. In evolution, viruses are an important means of horizontal gene transfer, which increases genetic diversity.

Opinions differ on whether viruses are a form of life or organic structures that interact with living organisms. They are considered by some to be a life form, because they carry genetic material, reproduce by creating multiple copies of themselves through self-assembly, and evolve through natural selection. However they lack key characteristics such as a cellular structure generally considered necessary to count as life. Because they possess some but not all such qualities, viruses have been described as replicators and as "organisms at the edge of life".

In terms of individual counts, tailed phage are the most abundant biological entities in the sea.

Bacteriophages, often just called phages, are viruses that parasite bacteria and archaea. Marine phages parasite marine bacteria and archaea, such as cyanobacteria. They are a common and diverse group of viruses and are the most abundant biological entity in marine environments, because their hosts, bacteria, are typically the numerically dominant cellular life in the sea. Generally there are about 1 million to 10 million viruses in each mL of seawater, or about ten times more double-stranded DNA viruses than there are cellular organisms, although estimates of viral abundance in seawater can vary over a wide range. Tailed bacteriophages appear to dominate marine ecosystems in number and diversity of organisms. Bacteriophages belonging to the families Corticoviridae, Inoviridae and Microviridae are also known to infect diverse marine bacteria.

Microorganisms make up about 70% of the marine biomass. It is estimated viruses kill 20% of this biomass each day and that there are 15 times as many viruses in the oceans as there are bacteria and archaea. Viruses are the main agents responsible for the rapid destruction of harmful algal blooms, which often kill other marine life. The number of viruses in the oceans decreases further offshore and deeper into the water, where there are fewer host organisms.

There are also archaean viruses which replicate within archaea: these are double-stranded DNA viruses with unusual and sometimes unique shapes. These viruses have been studied in most detail in the thermophilic archaea, particularly the orders Sulfolobales and Thermoproteales.

Viruses are an important natural means of transferring genes between different species, which increases genetic diversity and drives evolution. It is thought that viruses played a central role in the early evolution, before the diversification of bacteria, archaea and eukaryotes, at the time of the last universal common ancestor of life on Earth. Viruses are still one of the largest reservoirs of unexplored genetic diversity on Earth.

Marine Bacteria

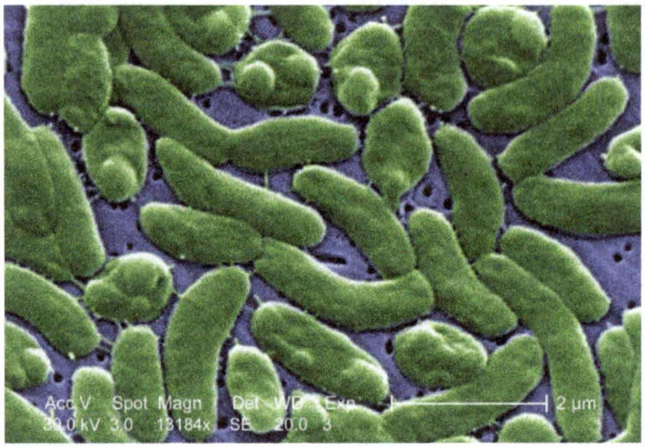

Vibrio vulnificus, a virulent bacterium found in estuaries and along coastal areas.

Bacteria constitute a large domain of prokaryotic microorganisms. Typically a few micrometres in length, bacteria have a number of shapes, ranging from spheres to rods and spirals. Bacteria were among the first life forms to appear on Earth, and are present in most of its habitats. Bacteria inhabit soil, water, acidic hot springs, radioactive waste, and the deep portions of Earth's crust. Bacteria also live in symbiotic and parasitic relationships with plants and animals.

Once regarded as plants constituting the class *Schizomycetes*, bacteria are now classified as prokaryotes. Unlike cells of animals and other eukaryotes, bacterial cells do not contain a nucleus and rarely harbour membrane-bound organelles. Although the term *bacteria* traditionally included all prokaryotes, the scientific classification changed after the discovery in the 1990s that prokaryotes consist of two very different groups of organisms that evolved from an ancient common ancestor. These evolutionary domains are called *Bacteria* and *Archaea*.

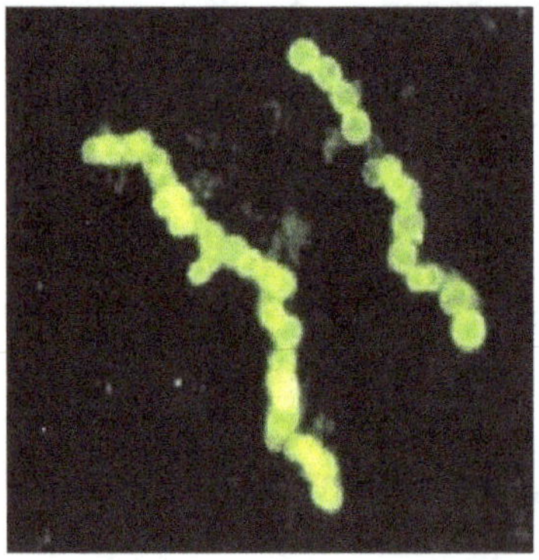

The marine Thiomargarita namibiensis, largest known bacterium.

The ancestors of modern bacteria were unicellular microorganisms that were the first forms of life to appear on Earth, about 4 billion years ago. For about 3 billion years, most organisms were microscopic, and bacteria and archaea were the dominant forms of life. Although bacterial fossils exist, such as stromatolites, their lack of distinctive morphology prevents them from being used to examine the history of bacterial evolution, or to date the time of origin of a particular bacterial species. However, gene sequences can be used to reconstruct the bacterial phylogeny, and these studies indicate that bacteria diverged first from the archaeal/eukaryotic lineage. Bacteria were also involved in the second great evolutionary divergence, that of the archaea and eukaryotes.

Here, eukaryotes resulted from the entering of ancient bacteria into endosymbiotic associations with the ancestors of eukaryotic cells, which were themselves possibly related to the Archaea. This involved the engulfment by proto-eukaryotic cells of alphaproteobacterial symbionts to form either mitochondria or hydrogenosomes, which are still found in all known Eukarya. Later on, some eukaryotes that already contained mitochondria also engulfed cyanobacterial-like organisms. This led to the formation of chloroplasts in algae and plants. There are also some algae that originated from even later endosymbiotic events. Here, eukaryotes engulfed a eukaryotic algae that developed into a "second-generation" plastid. This is known as secondary endosymbiosis.

The chloroplasts of glaucophytes have a peptidoglycan layer, evidence suggesting their endosymbiotic origin from cyanobacteria.

The largest known bacterium, the marine *Thiomargarita namibiensis*, can be visible to the naked eye and sometimes attains 0.75 mm (750 μm).

Marine Archaea

Archaea were initially viewed as extremophiles living in harsh environments, such as the yellow archaea pictured here in a hot spring, but they have since been found in a much broader range of habitats.

The archaea constitute a domain and kingdom of single-celled microorganisms. These microbes are prokaryotes, meaning they have no cell nucleus or any other membrane-bound organelles in their cells.

Archaea were initially classified as bacteria, but this classification is outdated. Archaeal cells have unique properties separating them from the other two domains of life, Bacteria and Eukaryota.

The Archaea are further divided into multiple recognized phyla. Classification is difficult because the majority have not been isolated in the laboratory and have only been detected by analysis of their nucleic acids in samples from their environment.

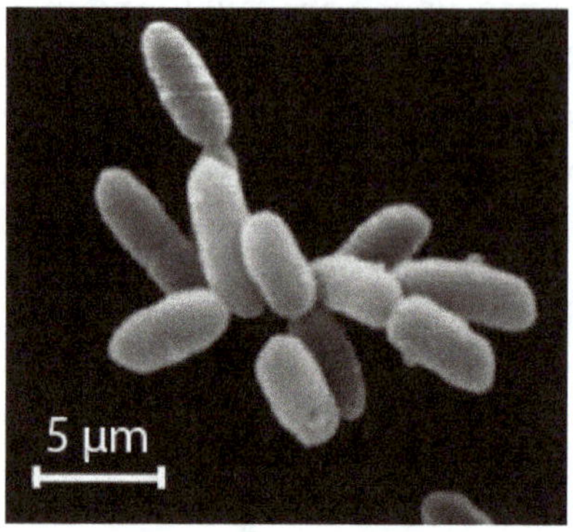

Halobacteria, found in water near saturated with salt, are now recognised as archaea.

Archaea and bacteria are generally similar in size and shape, although a few archaea have very strange shapes, such as the flat and square-shaped cells of *Haloquadratum walsbyi*. Despite this morphological similarity to bacteria, archaea possess genes and several metabolic pathways that are more closely related to those of eukaryotes, notably the enzymes involved in transcription and translation. Other aspects of archaeal biochemistry are unique, such as their reliance on ether lipids in their cell membranes, such as archaeols. Archaea use more energy sources than eukaryotes: these range from organic compounds, such as sugars, to ammonia, metal ions or even hydrogen gas. Salt-tolerant archaea (the Haloarchaea) use sunlight as an energy source, and other species of archaea fix carbon; however, unlike plants and cyanobacteria, no known species of archaea does both. Archaea reproduce asexually by binary fission, fragmentation, or budding; unlike bacteria and eukaryotes, no known species forms spores.

Archaea are particularly numerous in the oceans, and the archaea in plankton may be one of the most abundant groups of organisms on the planet. Archaea are a major part of Earth's life and may play roles in both the carbon cycle and the nitrogen cycle.

Flat, square-shaped cells of the archaea *Haloquadratum walsbyi*.

Marine Protists

Life originated as single-celled prokaryotes (bacteria and archaea) and later evolved into more complex eukaryotes. Eukaryotes are the more developed life forms known as plants, animals, fungi and protists. The term protist came into use historically as a term of convenience for eukaryotes that cannot be strictly classified as plants, animals or fungi. They are not a part of modern cladistics, because they are paraphyletic (lacking a common ancestor). Protists can be broadly divided into four groups depending on whether their nutrition is plant-like, animal-like, fungus-like, or a mixture of these.

Protists according to how they get food					
Type of protist		Description	Example		Other examples
Plant-like	Algae	Autotrophic protists that make their own food without needing to consume other organisms, usually by using photosynthesis	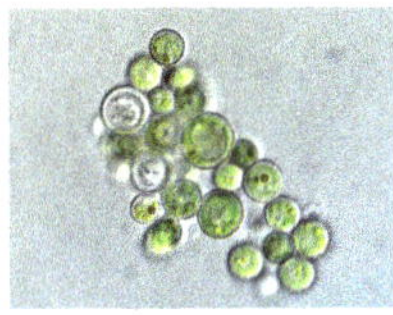	Red algae, Cyanidium sp.	Green algae, brown algae, diatoms and some dinoflagellates. Plant-like protists are important components of phytoplankton
Animal-like	Protozoans	Heterotrophic protists that get their food consuming other organisms	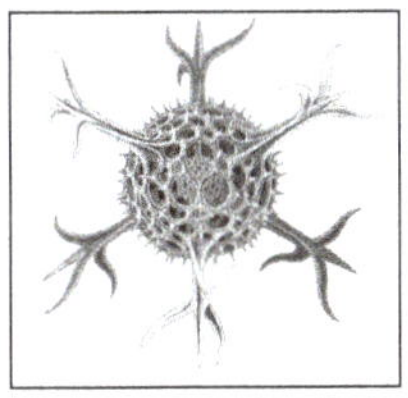	Radiolarian protist as drawn by Haeckel	Foraminiferans and some marine amoebae, ciliates and flagellates
Fungus-like	Slime moulds and slime nets	Saprotrophic protists that get their food from the remains of organisms that have broken down and decayed	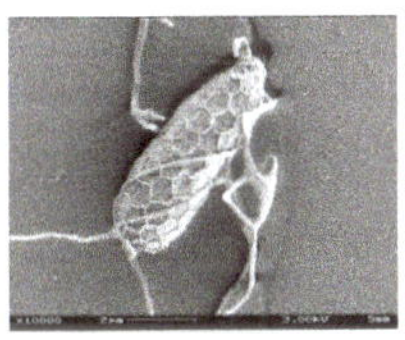	Marine slime nets form labyrinthine networks of tubes in which amoeba without pseudopods can travel	Marine lichen
Mixotropes	Various	Mixotrophic and osmotrophic protists that get their food from a combination of the above		Euglena mutabilis, a photosynthetic flagellate	Many marine mixotropes are found among protists, including among ciliates, Rhizaria and dinoflagellates

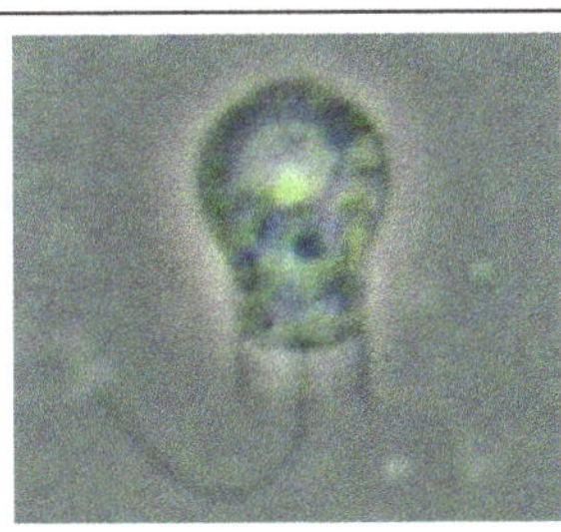	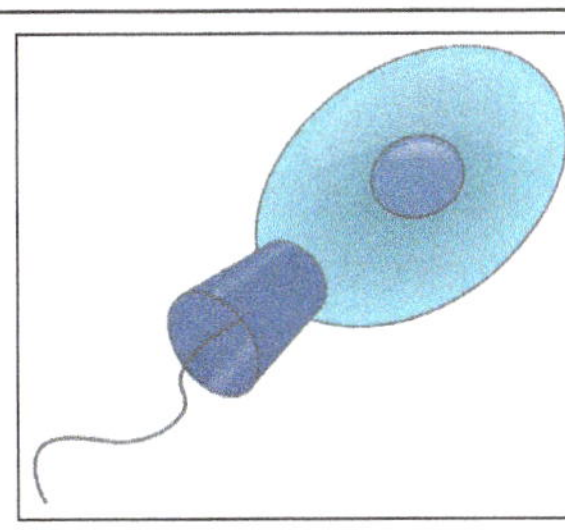
Micrograph	Cell schematic
Choanoflagellates, unicellular "collared" flagellate protists, are thought to be the closest living relatives of the animals.	

Protists are highly diverse organisms currently organised into 18 phyla, but are not easy to classify.

Studies have shown high protist diversity exists in oceans, deep sea-vents and river sediments, suggesting a large number of eukaryotic microbial communities have yet to be discovered. There has been little research on mixotrophic protists, but recent studies in marine environments found mixotrophic protests contribute a significant part of the protist biomass.

Single-celled and Microscopic Protists

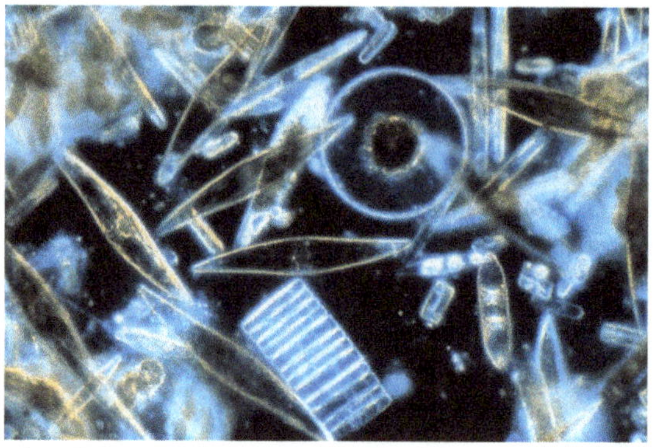

Diatoms are a major algae group generating about 20% of world oxygen production.

In contrast to the cells of prokaryotes, the cells of eukaryotes are highly organised. Plants, animals and fungi are usually multi-celled and are typically macroscopic. Most protists are single-celled and microscopic. But there are exceptions. Some single-celled marine protists are macroscopic. Some marine slime molds have unique life cycles that involve switching between unicellular, colonial, and multicellular forms. Other marine protist are neither single-celled nor microscopic, such as seaweed.

Giant kelp, a brown algae, is not a true plant, yet it is multicellular and can grow to 50m.

Protists have been described as a taxonomic grab bag where anything that doesn't fit into one of the main biological kingdoms can be placed. Some modern authors prefer to exclude multicellular organisms from the traditional definition of a protist, restricting protists to unicellular organisms. This more constrained definition excludes many brown, multicellular red and green algae, and slime molds.

Marine Microanimals

As juveniles, animals develop from microscopic stages, which can include spores, eggs and larvae. At least one microscopic animal group, the parasitic cnidarian Myxozoa, is unicellular in its adult form, and includes marine species. Other adult marine microanimals are multicellular. Microscopic adult arthropods are more commonly found inland in freshwater, but there are marine species as well. Microscopic adult marine crustaceans include some copepods, cladocera and tardigrades (water bears). Some marine nematodes and rotifers are also too small to be recognised with the naked eye, as are many loricifera, including the recently discovered anaerobic species that spend their lives in an anoxic environment. Copepods contribute more to the secondary productivity and carbon sink of the world oceans than any other group of organisms.

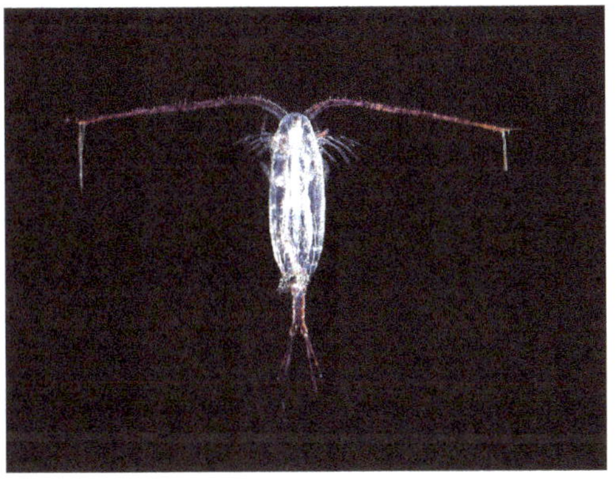

Over 10,000 marine species are copepods, small, often microscopic crustaceans.

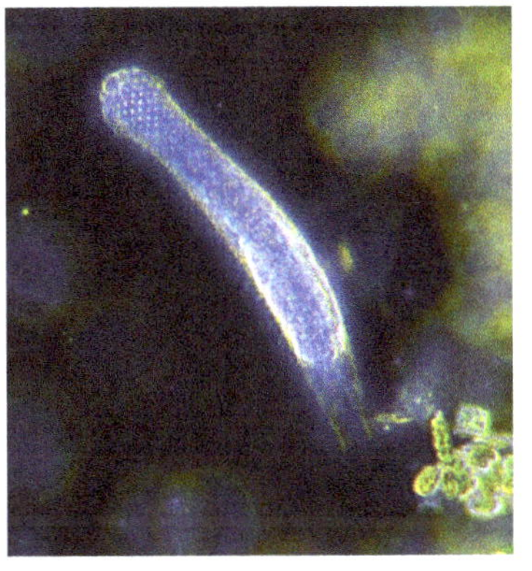

Darkfield photo of a gastrotrich, a worm-like animal living between sediment particles.

Fungi

Over 1500 species of fungi are known from marine environments. These are parasitic on marine algae or animals, or are saprobes feeding on dead organic matter from algae, corals, protozoan cysts, sea grasses, wood and other substrata. Spores of many species have special appendages which facilitate attachment to the substratum. They can also be found in sea foam. A diverse range of unusual secondary metabolites is produced by marine fungi.

Lichen on a rock in a marine splash zone. Lichens are mutualistic associations between a fungus and an alga or cyanobacterium.

Mycoplankton are saprotropic members of the plankton communities of marine and freshwater ecosystems. They are composed of filamentous free-living fungi and yeasts associated with planktonic particles or phytoplankton. Similar to bacterioplankton, these aquatic fungi play a significant role in heterotrophic mineralization and nutrient cycling. Mycoplankton can be up to 20 mm in diameter and over 50 mm in length.

A typical milliliter of seawater contains about 10^3 to 10^4 fungal cells. This number is greater in coastal ecosystems and estuaries due to nutritional runoff from terrestrial communities. A higher diversity of mycoplankton is found around coasts and in surface waters down to 1000 metres, with a vertical profile that depends on how abundant phytoplankton is. This profile changes between seasons due to changes in nutrient availability. Marine fungi survive in a constant oxygen deficient environment, and therefore depend on oxygen diffusion by turbulence and oxygen generated by photosynthetic organisms.

Marine fungi can be classified as:

- Lower fungi - Adapted to marine habitats (zoosporic fungi, including mastigomycetes: oomycetes and chytridiomycetes).
- Higher fungi - Filamentous, modified to planktonic lifestyle (hyphomycetes, ascomycetes, basidiomycetes). Most mycoplankton species are higher fungi.

Lichens are mutualistic associations between a fungus, usually an ascomycete, and an alga or a cyanobacterium. Several lichens are found in marine environments. Many more occur in the splash zone, where they occupy different vertical zones depending on how tolerant they are to submersion. Some lichens live a long time; one species has been dated at 8,600 years. However their lifespan is difficult to measure because what defines the same lichen is not precise. Lichens grow by vegetatively breaking off a piece, which may or may not be defined as the same lichen, and two lichens of different ages can merge, raising the issue of whether it is the same lichen.

The sea snail *Littoraria irrorata* damages plants of *Spartina* in the sea marshes where it lives, which enables spores of intertidal ascomycetous fungi to colonise the plant. The snail then eats the fungal growth in preference to the grass itself.

According to fossil records, fungi date back to the late Proterozoic era 900-570 million years ago. Fossil marine lichens 600 million years old have been discovered in China. It has been hypothesized that mycoplankton evolved from terrestrial fungi, likely in the Paleozoic era (390 million years ago).

Origin of Animals

The earliest animals were marine invertebrates, that is, vertebrates came later. Animals are multicellular eukaryotes, and are distinguished from plants, algae, and fungi by lacking cell walls. Marine invertebrates are animals that inhabit a marine environment apart from the vertebrate members of the chordate phylum; invertebrates lack a vertebral column. Some have evolved a shell or a hard exoskeleton.

The earliest animal fossils may belong to the genus *Dickinsonia*, 571 million to 541 million years ago. Individual *Dickinsonia* typically resembles a bilaterally symmetrical ribbed oval. They kept growing until they were covered with sediment or otherwise killed, and spent most of their lives with their bodies firmly anchored to the sediment. Their taxonomic affinities are presently unknown, but their mode of growth is consistent with a bilaterian affinity.

Dickinsonia may be the earliest animal. They appear in the fossil record 571 million to 541 million years ago.

The earliest animal fossils may belong to the genus *Dickinsonia*, 571 million to 541 million years ago. Individual *Dickinsonia* typically resembles a bilaterally symmetrical ribbed oval. They kept growing until they were covered with sediment or otherwise killed, and spent most of their lives with their bodies firmly anchored to the sediment. Their taxonomic affinities are presently unknown, but their mode of growth is consistent with a bilaterian affinity.

Apart from *Dickinsonia*, the earliest widely accepted animal fossils are the rather modern-looking cnidarians (the group that includes coral, jellyfish, sea anemones and *Hydra*), possibly from around 580 Ma The Ediacara biota, which flourished for the last 40 million years before the start of the Cambrian, were the first animals more than a very few centimetres long. Like *Dickinsonia*, many were flat with a "quilted" appearance, and seemed so strange that there was a proposal to classify them as a separate kingdom, Vendozoa. Others, however, have been interpreted as early molluscs (*Kimberella*), echinoderms (*Arkarua*), and arthropods (*Spriggina*, *Parvancorina*). There is still debate about the classification of these specimens, mainly because the diagnostic features which allow taxonomists to classify more recent organisms, such as similarities to living organisms, are generally absent in the Ediacarans. However, there seems little doubt that *Kimberella* was at least a triploblastic bilaterian animal, in other words, an animal significantly more complex than the cnidarians.

Small shelly fauna are a very mixed collection of fossils found between the Late Ediacaran and Middle Cambrian periods. The earliest, *Cloudina*, shows signs of successful defense against predation and may indicate the start of an evolutionary arms race. Some tiny Early Cambrian shells almost certainly belonged to molluscs, while the owners of some "armor plates," *Halkieria* and *Microdictyon*, were eventually identified when more complete specimens were found in Cambrian lagerstätten that preserved soft-bodied animals.

Body Plans and Phyla

Invertebrates are grouped into different phyla. Informally phyla can be thought of as a way of grouping organisms according to their body plan. A body plan refers to a blueprint which describes the shape or morphology of an organism, such as its symmetry, segmentation and the disposition of its appendages. The idea of body plans originated with vertebrates, which were grouped into one phylum. But the vertebrate body plan is only one of many, and invertebrates consist of many phyla or body plans. The history of the discovery of body plans can be seen as a movement from a worldview centred on vertebrates, to seeing the vertebrates as one body plan among many. Among the pioneering zoologists, Linnaeus identified two body plans outside the vertebrates; Cuvier identified three; and Haeckel had four, as well as the Protista with eight more, for a total of twelve. For comparison, the number of phyla recognised by modern zoologists has risen to 35.

Kimberella, an early mollusc important for understanding the Cambrian explosion. Invertebrates are grouped into different phyla (body plans).

Historically body plans were thought of as having evolved rapidly during the Cambrian explosion, but a more nuanced understanding of animal evolution suggests a gradual development of body plans throughout the early Palaeozoic and beyond. More generally a phylum can be defined in two ways: as described above, as a group of organisms with a certain degree of morphological or developmental similarity (the phenetic definition), or a group of organisms with a certain degree of evolutionary relatedness (the phylogenetic definition).

Opabinia, an extinct stem group arthropod appeared in the Middle Cambrian.

In the 1970s there was already a debate about whether the emergence of the modern phyla was "explosive" or gradual but hidden by the shortage of Precambrian animal fossils. A re-analysis of fossils from the Burgess Shale lagerstätte increased interest in the issue when it revealed animals, such as *Opabinia*, which did not fit into any known phylum. At the time these were interpreted as evidence that the modern phyla had evolved very rapidly in the Cambrian explosion and that the Burgess Shale's "weird wonders" showed that the Early Cambrian was a uniquely experimental period of animal evolution. Later discoveries of similar animals and the development of new theoretical approaches led to the conclusion that many of the "weird wonders" were evolutionary "aunts" or "cousins" of modern groups—for example that *Opabinia* was a member of the lobopods, a group which includes the ancestors of the arthropods, and that it may have been closely related to the modern tardigrades. Nevertheless, there is still much debate about whether the Cambrian explosion was really explosive and, if so, how and why it happened and why it appears unique in the history of animals.

Basal Invertebrate Animals

The most basal animal phyla, the animals that evolved first, are the Porifera, Ctenophora, Placozoa and Cnidaria. None of these basal body plans exhibit bilateral symmetry.

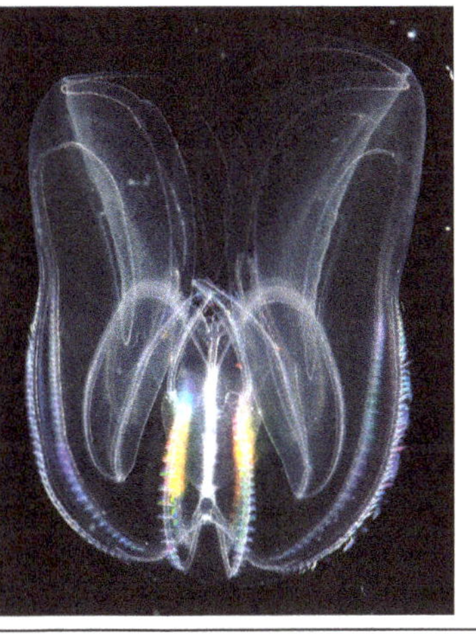

There has been much controversy over which invertebrate phyla, sponges or comb jellies, is the most basal. Currently sponges seem to be getting the upper hand.

Marine Sponges

Sponges are animals of the phylum Porifera. They are multicellular organisms that have bodies full of pores and channels allowing water to circulate through them, consisting of jelly-like mesohyl sandwiched between two thin layers of cells. They have unspecialized cells that can transform into other types and that often migrate between the main cell layers and the mesohyl in the process. Sponges do not have nervous, digestive or circulatory systems. Instead, most rely on maintaining a constant water flow through their bodies to obtain food and oxygen and to remove wastes.

Sponges are similar to other animals in that they are multicellular, heterotrophic, lack cell walls and produce sperm cells. Unlike other animals, they lack true tissues and organs, and have no body symmetry. The shapes of their bodies are adapted for maximal efficiency of water flow through the central cavity, where it deposits nutrients, and leaves through a hole called the osculum. Many sponges have internal skeletons of spongin and/or spicules of calcium carbonate or silicon dioxide. All sponges are sessile aquatic animals. Although there are freshwater species, the great majority are marine (salt water) species, ranging from tidal zones to depths exceeding 8,800 m (5.5 mi). Some sponges live to great ages; there is evidence of the deep-sea glass sponge *Monorhaphis chuni* living about 11,000 years.

While most of the approximately 5,000–10,000 known species feed on bacteria and other food particles in the water, some host photosynthesizing micro-organisms as endosymbionts and these alliances often produce more food and oxygen than they consume. A few species of sponge that live in food-poor environments have become carnivores that prey mainly on small crustaceans.

Sponges are perhaps the most basal animals. They have no nervous, digestive or circulatory system.

Linnaeus mistakenly identified sponges as plants in the order Algae. For a long time thereafter sponges were assigned to a separate subkingdom, Parazoa (meaning beside the animals). They are now classified as a paraphyletic phylum from which the higher animals have evolved.

Ctenophores

Together with sponges, brilliantly bioluminescent ctenophores (comb jellies) are the most basal animals.

Ctenophores, commonly known as comb jellies, are a phylum that live worldwide in marine waters. They are the largest non-colonial animals to swim with the help of cilia (hairs or combs). Coastal species need to be tough enough to withstand waves and swirling sediment, but some oceanic species are so fragile and transparent that it is very difficult to capture them intact for study. In the past ctenophores were thought to have only a modest presence in the ocean, but it is now known they are often significant and even dominant parts of the planktonic biomass.

The phylum has about 150 known species with a wide range of body forms. Sizes range from a few millimeters to 1.5 m (4 ft 11 in). Cydippids are egg-shaped with their cilia arranged in eight radial comb rows, and deploy retractable tentacles for capturing prey. The benthic platyctenids are generally combless and flat. The coastal beroids have gaping mouths and lack tentacles. Most adult ctenophores prey on microscopic larvae and rotifers and small crustaceans but beroids prey on other ctenophores.

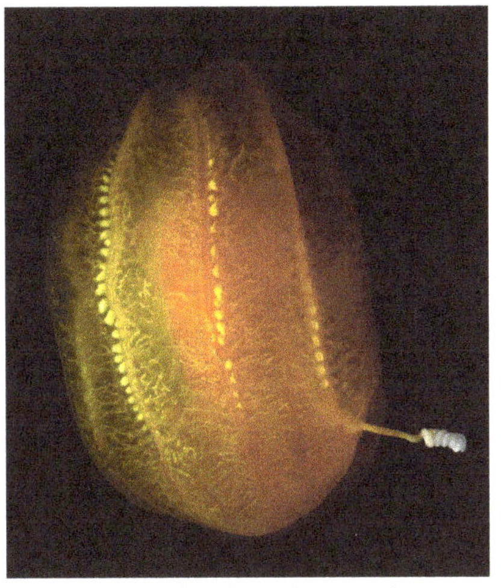

Egg-shaped cydippid ctenophore.

Early writers combined ctenophores with cnidarians. Ctenophores resemble cnidarians in relying on water flow through the body cavity for both digestion and respiration, as well as in having a decentralized nerve net rather than a brain. Also like cnidarians, the bodies of ctenophores consist of a mass of jelly, with one layer of cells on the outside and another lining the internal cavity. In ctenophores, however, these layers are two cells deep, while those in cnidarians are only a single cell deep. While cnidarians exhibit radial symmetry, ctenophores have two anal canals which exhibit biradial symmetry (half-turn rotational symmetry). The position of the ctenophores in the evolutionary family tree of animals has long been debated, and the majority view at present, based on molecular phylogenetics, is that cnidarians and bilaterians are more closely related to each other than either is to ctenophores.

Placozoa

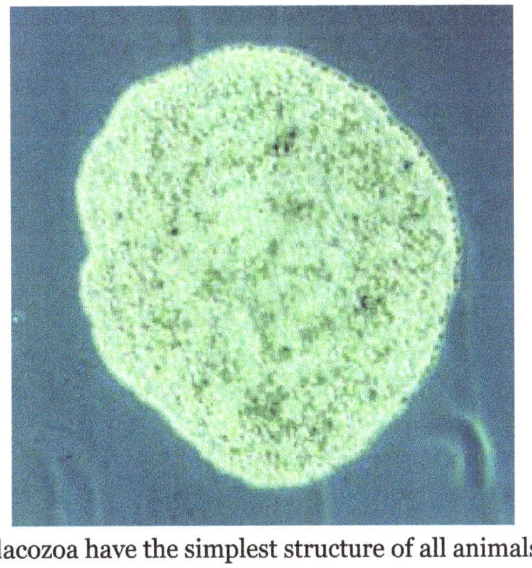

Placozoa have the simplest structure of all animals.

Placozoa have the simplest structure of all animals. They are a basal form of free-living (non-parasitic) multicellular organism. A common name does not yet exist for them. They form a phylum containing, to date, only three species. The first species was discovered in 1883, the classical *Trichoplax adhaerens*. Two more species have been discovered since 2017.

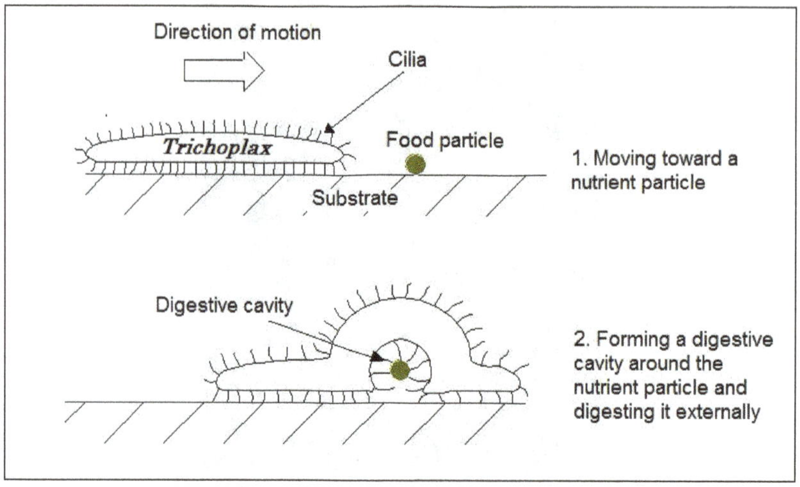

Food uptake by *Trichoplax adhaerens.*

Trichoplax is a small, flattened, animal about one mm across and usually about 25 μm thick. Like the amoebae they superficially resemble, they continually change their external shape. In addition, spherical phases occasionally form which may facilitate movement. *Trichoplax* lacks tissues and organs. There is no manifest body symmetry, so it is not possible to distinguish anterior from posterior or left from right. It is made up of a few thousand cells of six types in three distinct layers. The outer layer of simple epithelial cells bear cilia which the animal uses to help it creep along the seafloor. *Trichoplax* feed by engulfing and absorbing food particles – mainly microbes and organic detritus – with their underside.

Marine Cnidarians

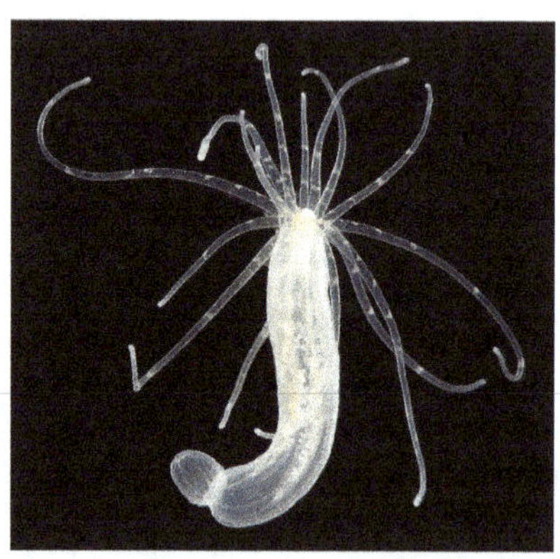

Cnidarians, like this starlet sea anemone, are the simplest animals to organise cells into tissue. Yet they have the same genes that form the vertebrate (including human) head.

Cnidarians are distinguished by the presence of stinging cells, specialized cells that they use mainly for capturing prey. Cnidarians include corals, sea anemones, jellyfish and hydrozoans. They form a phylum containing over 10,000 species of animals found exclusively in aquatic (mainly marine) environments. Their bodies consist of mesoglea, a non-living jelly-like substance, sandwiched between two layers of epithelium that are mostly one cell thick. They have two basic body forms: swimming medusae and sessile polyps, both of which are radially symmetrical with mouths surrounded by tentacles that bear cnidocytes. Both forms have a single orifice and body cavity that are used for digestion and respiration.

Fossil cnidarians have been found in rocks formed about 580 million years ago. Fossils of cnidarians that do not build mineralized structures are rare. Scientists currently think cnidarians, ctenophores and bilaterians are more closely related to calcareous sponges than these are to other sponges, and that anthozoans are the evolutionary "aunts" or "sisters" of other cnidarians, and the most closely related to bilaterians.

Cnidarians are the simplest animals in which the cells are organised into tissues. The starlet sea anemone is used as a model organism in research. It is easy to care for in the laboratory and a protocol has been developed which can yield large numbers of embryos on a daily basis. There is a remarkable degree of similarity in the gene sequence conservation and complexity between the sea anemone and vertebrates. In particular, genes concerned in the formation of the head in vertebrates are also present in the anemone.

Their tentacles sting and paralyse small fish.

Bilateral Invertebrate Animals

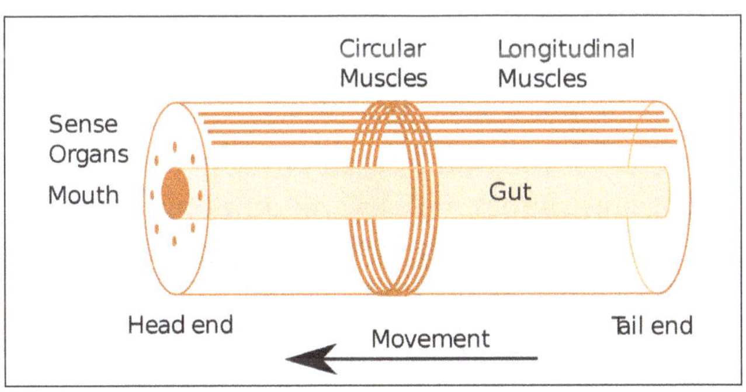

The above image shows Idealised wormlike bilaterian body plan. With a cylindrical body and a direction of movement the animal has head and tail ends. Sense organs and mouth form the basis of the head. Opposed circular and longitudinal muscles enable peristaltic motion.

Some of the earliest bilaterians were wormlike, and a bilaterian body can be conceptualized as a cylinder with a gut running between two openings, the mouth and the anus. Around the gut it has an internal body cavity, a coelom or pseudocoelom. Animals with this bilaterally symmetric body plan have a head (anterior) end and a tail (posterior) end as well as a back (dorsal) and a belly (ventral); therefore they also have a left side and a right side.

Having a front end means that this part of the body encounters stimuli, such as food, favouring cephalisation, the development of a head with sense organs and a mouth. The body stretches back from the head, and many bilaterians have a combination of circular muscles that constrict the body, making it longer, and an opposing set of longitudinal muscles, that shorten the body; these enable soft-bodied animals with a hydrostatic skeleton to move by peristalsis. They also have a gut that extends through the basically cylindrical body from mouth to anus. Many bilaterian phyla have primary larvae which swim with cilia and have an apical organ containing sensory cells. However, there are exceptions to each of these characteristics; for example, adult echinoderms are radially symmetric (unlike their larvae), and certain parasitic worms have extremely simplified body structures. The original bilaterian may have been a bottom dwelling worm with a single body opening.

Protostomes

Protostomes are a superphylum of animals. It is a sister clade of the deuterostomes, with which it forms the Nephrozoa clade. Protostomes are distinguished from deuterostomes by the way their embryos develop. In protostomes the first opening that develops becomes the mouth, while in deuterostomes it becomes the anus.

Marine Worms

Many marine worms are related only distantly, so they form a number of different phyla. The worm shown is an arrow worm, found worldwide as a predatory component of plankton.

Worms (Old English for *serpents*) form a number of phylums. Different groups of marine worms are related only distantly, so they are found in several different phyla such as the Annelida (segmented worms), Chaetognatha (arrow worms), Phoronida (horseshoe worms), and Hemichordata. All worms, apart from the Hemichordata, are protostomes.

The typical body plan of a worm involves long cylindrical tube-like bodies and no limbs. Marine worms vary in size from microscopic to over 1 metre (3.3 ft) in length for some marine polychaete worms (bristle worms) and up to 58 metres (190 ft) for the marine nemertean worm (bootlace worm). Some marine worms occupy a small variety of parasitic niches, living inside the bodies of other animals, while others live more freely in the marine environment or by burrowing underground. Many of these worms have specialized tentacles used for exchanging oxygen and carbon dioxide and also may be used for reproduction. Some marine worms are tube worms, such as the giant tube worm which lives in waters near underwater volcanoes and can withstand temperatures up to 90 degrees Celsius. Platyhelminthes (flatworms) form another worm phylum which includes a class of parasitic tapeworms. The marine tapeworm *Polygonoporus giganticus*, found in the gut of sperm whales, can grow to over 30 m (100 ft).

Nematodes (roundworms) constitute a further worm phylum with tubular digestive systems and an opening at both ends. Over 25,000 nematode species have been described, of which more than half are parasitic. It has been estimated another million remain undescribed. They are ubiquitous in marine, freshwater and terrestrial environments, where they often outnumber other animals in both individual and species counts. They are found in every part of the earth's lithosphere, from the top of mountains to the bottom of oceanic trenches. By count they represent 90% of all animals on the ocean floor. Their numerical dominance, often exceeding a million individuals per square meter and accounting for about 80% of all individual animals on earth, their diversity of life cycles, and their presence at various trophic levels point at an important role in many ecosystems.

Bloodworms are typically found on the bottom of shallow marine waters.

Marine Molluscs

Bolinus brandaris a sea snail from which the Phoenicians extracted royal Tyrian red dye.

Molluscs form a phylum with about 85,000 extant recognized species. They are the largest marine phylum in terms of species count, containing about 23% of all the named marine organisms. Molluscs have more varied forms than other invertebrate phylums. They are highly diverse, not just in size and in anatomical structure, but also in behaviour and in habitat.

The mollusc phylum is divided into 9 or 10 taxonomic classes. These classes include gastropods, bivalves and cephalopods, as well as other lesser-known but distinctive classes. Gastropods with protective shells are referred to as snails, whereas gastropods without protective shells are referred to as slugs. Gastropods are by far the most numerous molluscs in terms of species. Bivalves include clams, oysters, cockles, mussels, scallops, and numerous other families. There are about 8,000 marine bivalves species (including brackish water and estuarine species). A deep sea ocean quahog clam has been reported as having lived 507 years making it the longest recorded life of all animals apart from colonial animals, or near-colonial animals like sponges.

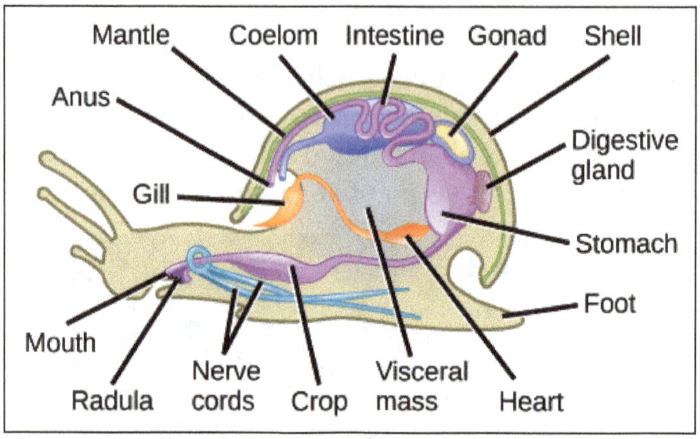

Bigfin reef squid displaying vivid iridescence at night. Cephalopods are the most neurologically advanced invertebrates.

Cephalopods include octopus, squid and cuttlefish. About 800 living species of marine cephalopods have been identified, and an estimated 11,000 extinct taxa have been described. They are found in all oceans, but there are no fully freshwater cephalopods.

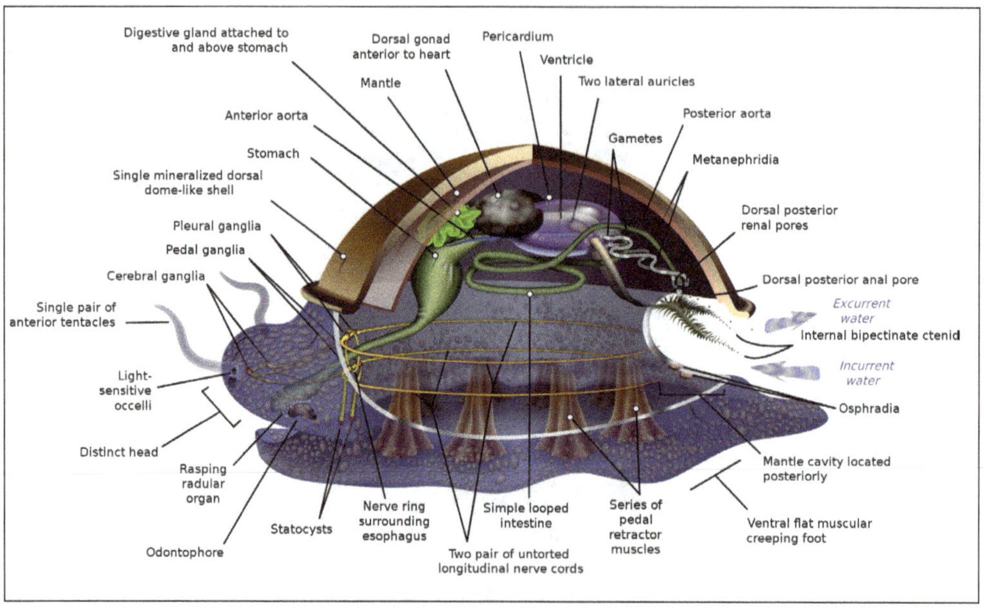

Hypothetical ancestral mollusc.

Molluscs have such diverse shapes that many textbooks base their descriptions of molluscan anatomy on a generalized or *hypothetical ancestral mollusc*. This generalized mollusc is unsegmented and bilaterally symmetrical with an underside consisting of a single muscular foot. Beyond that it has three further key features. Firstly, it has a muscular cloak called a mantle covering its viscera and containing a significant cavity used for breathing and excretion. A shell secreted by the mantle covers the upper surface. Secondly (apart from bivalves) it has a rasping tongue called a radula used for feeding. Thirdly, it has a nervous system including a complex digestive system using microscopic, muscle-powered hairs called cilia to exude mucus. The generalized mollusc has two paired nerve cords (three in bivalves). The brain, in species that have one, encircles the esophagus. Most molluscs have eyes and all have sensors detecting chemicals, vibrations, and touch.

Good evidence exists for the appearance of marine gastropods, cephalopods and bivalves in the Cambrian period 541 to 485.4 million years ago.

Marine Arthropods

Pneumodesmus newmani is the first known invertebrate to colonise land. It lived in the Early Devonian.

Arthropods have an exoskeleton (external skeleton), a segmented body, and jointed appendages (paired appendages). They form a phylum which includes insects, arachnids, myriapods, and crustaceans. Arthropods are characterized by their jointed limbs and cuticle made of chitin, often mineralised with calcium carbonate. The arthropod body plan consists of segments, each with a pair of appendages. The rigid cuticle inhibits growth, so arthropods replace it periodically by moulting. Their versatility has enabled them to become the most species-rich members of all ecological guilds in most environments.

The evolutionary ancestry of arthropods dates back to the Cambrian period and is generally regarded as monophyletic. However, basal relationships of arthropods with extinct phyla such as lobopodians have recently been debated.

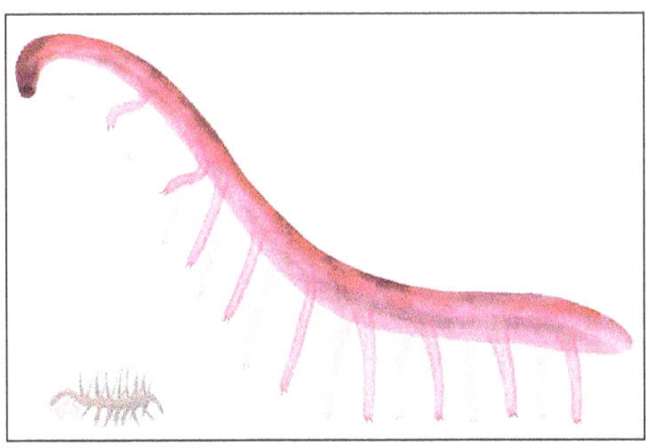

Some palaeontologists think Lobopodia represents a basal grade which lead to an arthropod body plan.

Extant marine arthropods range in size from the microscopic crustacean *Stygotantulus* to the Japanese spider crab. Arthropods' primary internal cavity is a hemocoel, which accommodates their internal organs, and through which their haemolymph - analogue of blood - circulates; they have open circulatory systems. Like their exteriors, the internal organs of arthropods are generally built of repeated segments. Their nervous system is "ladder-like", with paired ventral nerve cords running through all segments and forming paired ganglia in each segment. Their heads are formed by fusion of varying numbers of segments, and their brains are formed by fusion of the ganglia of these segments and encircle the esophagus. The respiratory and excretory systems of arthropods vary, depending as much on their environment as on the subphylum to which they belong.

Arthropod vision relies on various combinations of compound eyes and pigment-pit ocelli: in most species the ocelli can only detect the direction from which light is coming, and the compound eyes are the main source of information. Arthropods also have a wide range of chemical and mechanical sensors, mostly based on modifications of the many setae (bristles) that project through their cuticles. Arthropod methods of reproduction are diverse: terrestrial species use some form of internal fertilization while marine species lay eggs using either internal or external fertilization. Arthropod hatchlings vary from miniature adults to grubs that lack jointed limbs and eventually undergo a total metamorphosis to produce the adult form.

Deuterostomes

In deuterostomes the first opening that develops in the growing embryo becomes the anus, while in protostomes it becomes the mouth. Deuterostomes form a superphylum of animals and are the sister clade of the protostomes. The earliest known deuterostomes are *Saccorhytus* fossils from about 540 million years ago. The *Saccorhytus* mouth may have functioned also as its anus.

Fossil showing the mouth of the earliest known deuterostome, 540 mya.

Echinoderms

Echinoderms is a phylum which contains only marine invertebrates. The phylum contains about 7000 living species, making it the second-largest grouping of deuterostomes, after the chordates.

Echinoderm literally means "spiny skin", as this water melon sea urchin illustrates.

Adult echinoderms are recognizable by their radial symmetry (usually five-point) and include starfish, sea urchins, sand dollars, and sea cucumbers, as well as the sea lilies. Echinoderms are found at every ocean depth, from the intertidal zone to the abyssal zone. They are unique among animals in having bilateral symmetry at the larval stage, but fivefold symmetry (pentamerism, a special type of radial symmetry) as adults.

Echinoderms are important both biologically and geologically. Biologically, there are few other groupings so abundant in the biotic desert of the deep sea, as well as shallower oceans. Most echinoderms are able to regenerate tissue, organs, limbs, and reproduce asexually; in some cases, they can undergo complete regeneration from a single limb. Geologically, the value of echinoderms is in their ossified skeletons, which are major contributors to many limestone formations, and can provide valuable clues as to the geological environment.

It is held by some scientists that the radiation of echinoderms was responsible for the Mesozoic Marine Revolution. Aside from the hard-to-classify *Arkarua* (a Precambrian animal with echinoderm-like pentamerous radial symmetry), the first definitive members of the phylum appeared near the start of the Cambrian.

Hemichordates

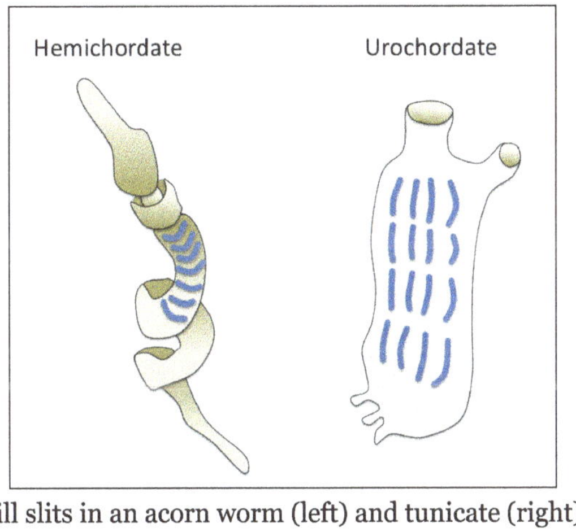

Gill slits in an acorn worm (left) and tunicate (right).

Hemichordates form a sister phylum to the echinoderms. They are solitary worm-shaped organisms rarely seen by humans because of their lifestyle. They include two main groups, the acorn worms and the Pterobranchia. Pterobranchia form a class containing about 30 species of small worm-shaped animals that live in secreted tubes on the ocean floor. Acorn worms form a class containing about 111 species that generally live in U-shaped burrows on the seabed, from the shoreline to a depth of 3000 metres. The worms lie there with the proboscis sticking out of one opening in the burrow, subsisting as deposit feeders or suspension feeders. It is supposed the ancestors of acorn worms used to live in tubes like their relatives, the Pterobranchia, but eventually started to live a safer and more sheltered existence in sediment burrows. Some of these worms may grow to be very long; one particular species may reach a length of 2.5 metres (8 ft 2 in), although most acorn worms are much smaller.

Acorn worms are more highly specialised and advanced than other worm-like organisms. They have a circulatory system with a heart that also functions as a kidney. Acorn worms have gill-like structures they use for breathing, similar to the gills of fish. Therefore, acorn worms are sometimes said to be a link between classical invertebrates and vertebrates. Acorn worms continually form new gill slits as they grow in size, and some older individuals have more than a hundred on each side. Each slit consists of a branchial chamber opening to the pharynx through a U-shaped cleft. Cilia push water through the slits, maintaining a constant flow, just as in fish. Some acorn worms also have a postanal tail which may be homologous to the post-anal tail of vertebrates.

The three-section body plan of the acorn worm is no longer present in the vertebrates, except in the anatomy of the frontal neural tube, later developed into a brain divided into three parts. This means some of the original anatomy of the early chordate ancestors is still present in vertebrates even if it is not always visible. One theory is the three-part body originated from an early common ancestor of the deuterostomes, and maybe even from a common bilateral ancestor of both deuterostomes and protostomes. Studies have shown the gene expression in the embryo share three of the same signaling centers that shape the brains of all vertebrates, but instead of taking part in the formation of their neural system, they are controlling the development of the different body regions.

Marine Chordates

The chordate phylum has three subphylums, one of which is the vertebrates. The other two subphylums are marine invertebrates: the tunicates (salps and sea squirts) and the cephalochordates (such as lancelets). Invertebrate chordates are close relatives to vertebrates. In particular, there has been discussion about how closely some extinct marine species, such as Pikaiidae, Palaeospondylus, Zhongxiniscus and Vetulicolia, might relate ancestrally to vertebrates.

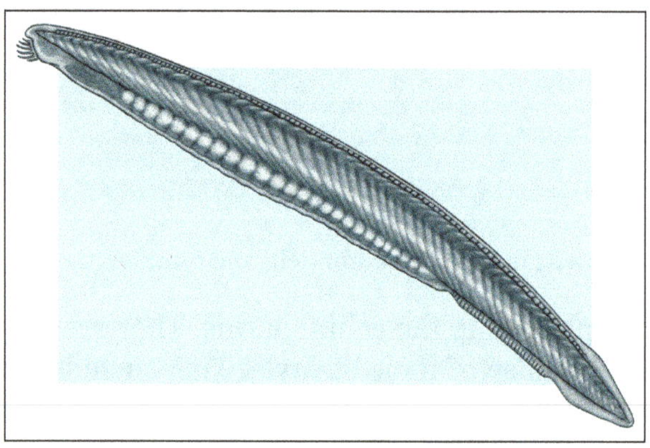

The lancelet, a small translucent fish-like cephalochordate, is the closest living invertebrate relative of the vertebrates.

Vertebrate Animals

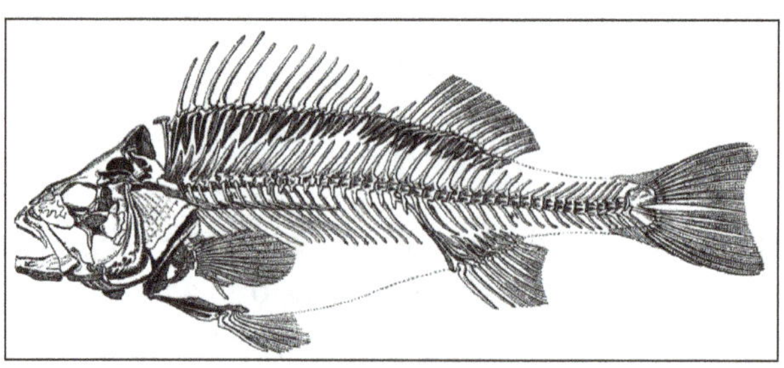

Ray-finned fish

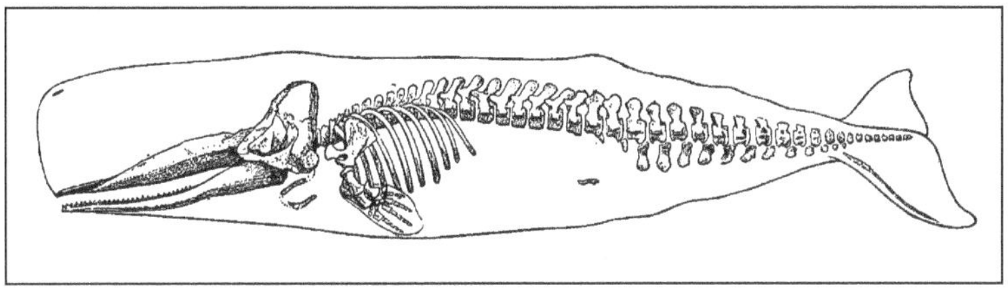

Marine tetrapod (sperm whale)

Skeletal structures showing the vertebral column and internal skeleton running from the head to the tail. Vertebrates are a subphylum of chordates. They are chordates that have a vertebral column (backbone). The vertebral column provides the central support structure for an internal skeleton which gives shape, support, and protection to the body and can provide a means of anchoring fins or limbs to the body. The vertebral column also serves to house and protect the spinal cord that lies within the vertebral column. Marine vertebrates can be divided into marine fish and marine tetrapods.

Marine Fish

Fish typically breathe by extracting oxygen from water through gills and have a skin protected by scales and mucous. They use fins to propel and stabilise themselves in the water, and usually have a two-chambered heart and eyes well adapted to seeing underwater, as well as other sensory systems. Over 33,000 species of fish have been described as of 2017, of which about 20,000 are marine fish.

Jawless Fish

Early fish had no jaws. Most went extinct, but two groups survived: hagfish and lampreys. Hagfish form a class of about 20 species of eel-shaped, slime-producing marine fish. They are the only known living animals that have a skull but no vertebral column. Lampreys form a superclass containing 38 known extant species of jawless fish. The adult lamprey is characterized by a toothed, funnel-like sucking mouth. Although they are well known for boring into the flesh of other fish to suck their blood, only 18 species of lampreys are actually parasitic. Together hagfish and lampreys are the sister group to vertebrates. Living hagfish remain similar to hagfish from around 300 million years ago. The lampreys are a very ancient lineage of vertebrates, though their exact relationship to hagfishes and jawed vertebrates is still a matter of dispute. Molecular analysis since 1992 has suggested that hagfish are most closely related to lampreys, and so also are vertebrates in a monophyletic sense. Others consider them a sister group of vertebrates in the common taxon of craniata.

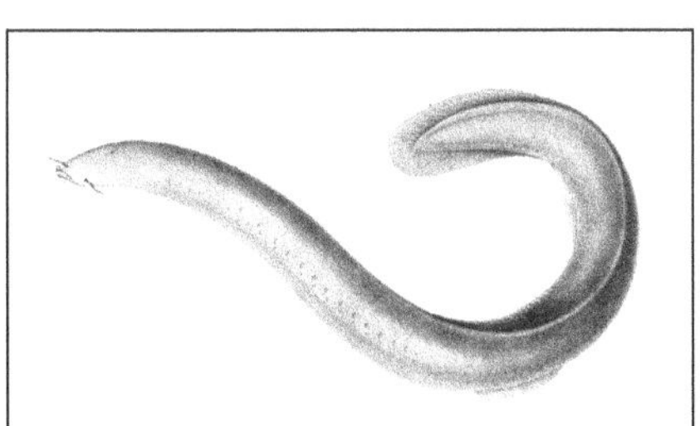

Hagfish are the only known living animals with a skull but no vertebral column.

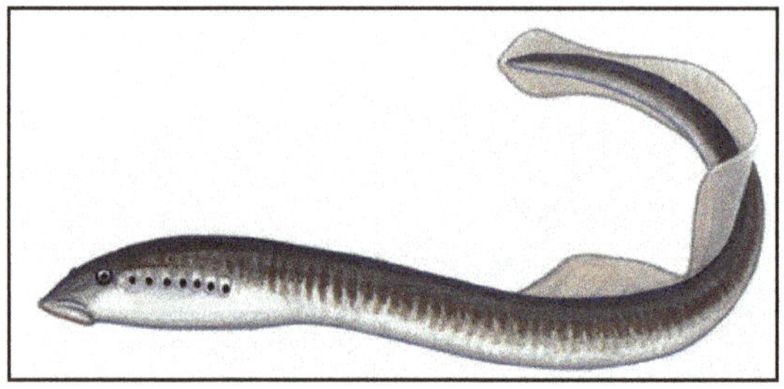

Lampreys are often parasitic and have a toothed, funnel-like sucking mouth.

Pteraspidomorphi is an extinct class of early jawless fish ancestral to jawed vertebrates. The few characteristics they share with the latter are now considered as primitive for all vertebrates.

Around the start of the Devonian, fish started appearing with a deep remodelling of the vertebrate skull that resulted in a jaw. All vertebrate jaws, including the human jaw, have evolved from these early fish jaws. The appearance of the early vertebrate jaw has been described as "perhaps the most profound and radical evolutionary step in vertebrate history". Jaws make it possible to capture, hold, and chew prey. Fish without jaws had more difficulty surviving than fish with jaws, and most jawless fish became extinct during the Triassic period.

Cartilaginous Fish

Jawed fish fall into two main groups: fish with bony internal skeletons and fish with cartilaginous internal skeletons. Cartilaginous fish, such as sharks and rays, have jaws and skeletons made of cartilage rather than bone. Megalodon is an extinct species of shark that lived about 28 to 1.5 Ma. It looked much like a stocky version of the great white shark, but was much larger with fossil lengths reaching 20.3 metres (67 ft). Found in all oceans it was one of the largest and most powerful predators in vertebrate history, and probably had a profound impact on marine life. The Greenland shark has the longest known lifespan of all vertebrates, about 400 years. Some sharks such as the great white are partially warm blooded and give live birth.

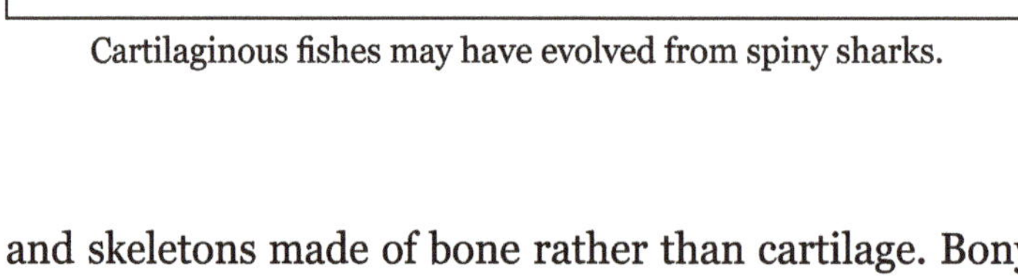

Cartilaginous fishes may have evolved from spiny sharks.

Bony Fish

Bony fish have jaws and skeletons made of bone rather than cartilage. Bony fish also have hard, bony plates called operculum which help them respire and protect their gills, and they often possess a swim bladder which they use for better control of their buoyancy. Bony fish can be further divided into those with lobe fins and those with ray fins. The approximate dates in the phylogenetic tree are from Near and Zhu.

Guiyu oneiros, the earliest-known bony fish lived during the Late Silurian 419 million years ago.

Lobe fins have the form of fleshy lobes supported by bony stalks which extend from the body. *Guiyu oneiros*, the earliest-known bony fish, lived during the Late Silurian 419 million years ago. It has the combination of both ray-finned and lobe-finned features, although analysis of the totality of its features place it closer to lobe-finned fish. Lobe fins evolved into the legs of the first tetrapod land vertebrates, so by extension an early ancestor of humans was a lobe-finned fish. Apart from the coelacanths and the lungfishes, lobe-finned fishes are now extinct. The remaining bony fish have ray fins. These are made of webs of skin supported by bony or horny spines (rays) which can be erected to control the fin stiffness.

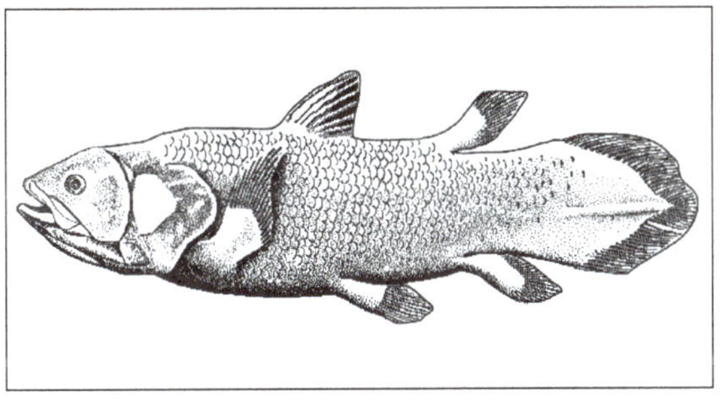

Lobe fins are bedded into the body by bony stalks. They evolved into the legs of the first tetrapod land vertebrates.

The main distinguishing feature of the chondrosteans (sturgeon, paddlefish, bichir and reedfish) is the cartilaginous nature of their skeletons. The ancestors of the chondrosteans are thought to be bony fish, but the characteristic of an ossified skeleton was lost in later evolutionary development, resulting in a lightening of the frame.

Neopterygians appeared sometime in the Late Permian, before dinosaurs. They were a very successful group of fish, because they could move more rapidly than their ancestors. Their scales and skeletons began to lighten during their evolution, and their jaws became more powerful and efficient.

Teleosts

About 96% of the world's fish species are teleosts, of which about 14,000 are marine species. Teleosts can be distinguished from other bony fish by their possession of a homocercal tail, a tail where the upper half mirrors the lower half. Another difference lies in their jaw bones – teleosts have modifications in the jaw musculature which make it possible for them to protrude their jaws. This enables them to grab prey and draw it into their mouth. In general, teleosts tend to be quicker and more flexible than more basal bony fishes. Their skeletal structure has evolved towards greater lightness. While teleost bones are well calcified, they are constructed from a scaffolding of struts, rather than the dense cancellous bones of holostean fish.

Teleosts have homocercal tails.

Teleosts are found in almost all marine habitats. They have enormous diversity, and range in size from adult gobies 8mm long to ocean sunfish weighting over 2,000 kg. The following images show something of the diversity in the shape and colour of modern marine teleosts.

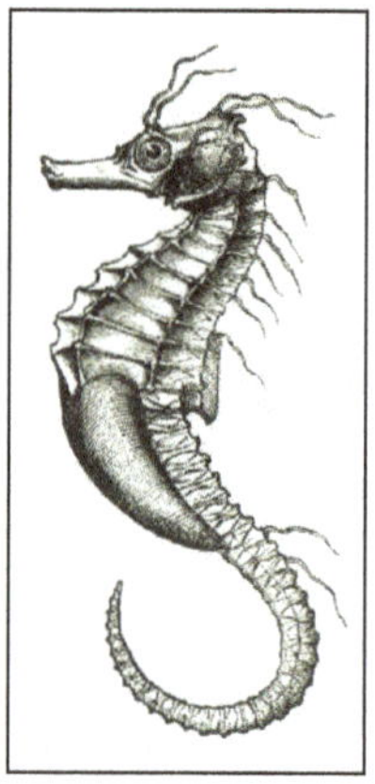

Seahorse

Marine Tetrapods

A tetrapod is a vertebrate with limbs (feet). Tetrapods evolved from ancient lobe-finned fishes about 400 million years ago during the Devonian Period when their earliest ancestors emerged from the sea and adapted to living on land. This change from a body plan for breathing and navigating in gravity-neutral water to a body plan with mechanisms enabling the animal to breath in air without dehydrating and move on land is one of the most profound evolutionary changes known. Tetrapods can be divided into four classes: amphibians, reptiles, birds and mammals.

Tiktaalik, an extinct lobe-finned fish, developed limb-like fins that could take it onto land.

Marine tetrapods are tetrapods that returned from land back to the sea again. The first returns to the ocean may have occurred as early as the Carboniferous Period whereas other returns occurred as recently as the Cenozoic, as in cetaceans, pinnipeds, and several modern amphibians. Amphibians live part of their life in water and part on land. They mostly require fresh water to reproduce. A few inhabit brackish water, but there are no true marine amphibians. There have been reports, however, of amphibians invading marine waters, such as a Black Sea invasion by the natural hybrid *Pelophylax esculentus* reported in 2010.

Reptiles

Marine iguana

Reptiles do not have an aquatic larval stage, and in this way are unlike amphibians. Most reptiles are oviparous, although several species of squamates are viviparous, as were some extinct aquatic clades—the fetus develops within the mother, contained in a placenta rather than an eggshell. As amniotes, reptile eggs are surrounded by membranes for protection and transport, which adapt them to reproduction on dry land. Many of the viviparous species feed their fetuses through various forms of placenta analogous to those of mammals, with some providing initial care for their hatchlings.

Some reptiles are more closely related to birds than other reptiles, and many scientists prefer to make Reptilia a monophyletic group which includes the birds. Extant non-avian reptiles which inhabit or frequent the sea include sea turtles, sea snakes, terrapins, the marine iguana, and the saltwater crocodile. Currently, of the approximately 12,000 extant reptile species and sub-species, only about 100 of are classed as marine reptiles.

Saltwater crocodile

Except for some sea snakes, most extant marine reptiles are oviparous and need to return to land to lay their eggs. Apart from sea turtles, the species usually spend most of their lives on or near land rather than in the ocean. Sea snakes generally prefer shallow waters nearby land, around islands, especially waters that are somewhat sheltered, as well as near estuaries. Unlike land snakes, sea snakes have evolved flattened tails which help them swim.

Some extinct marine reptiles, such as ichthyosaurs, evolved to be viviparous and had no requirement to return to land. Ichthyosaurs resembled dolphins. They first appeared about 245 million years ago and disappeared about 90 million years ago. The terrestrial ancestor of the ichthyosaur had no features already on its back or tail that might have helped along the evolutionary process. Yet the ichthyosaur developed a dorsal and tail fin which improved its ability to swim. The biologist Stephen Jay Gould said the ichthyosaur was his favourite example of convergent evolution. The earliest marine reptiles arose in the Permian. During the Mesozoic many groups of reptiles became adapted to life in the seas, including ichthyosaurs, plesiosaurs, mosasaurs, nothosaurs, placodonts, sea turtles, thalattosaurs and thalattosuchians. Marine reptiles were less numerous after mass extinction at the end of the Cretaceous.

Birds

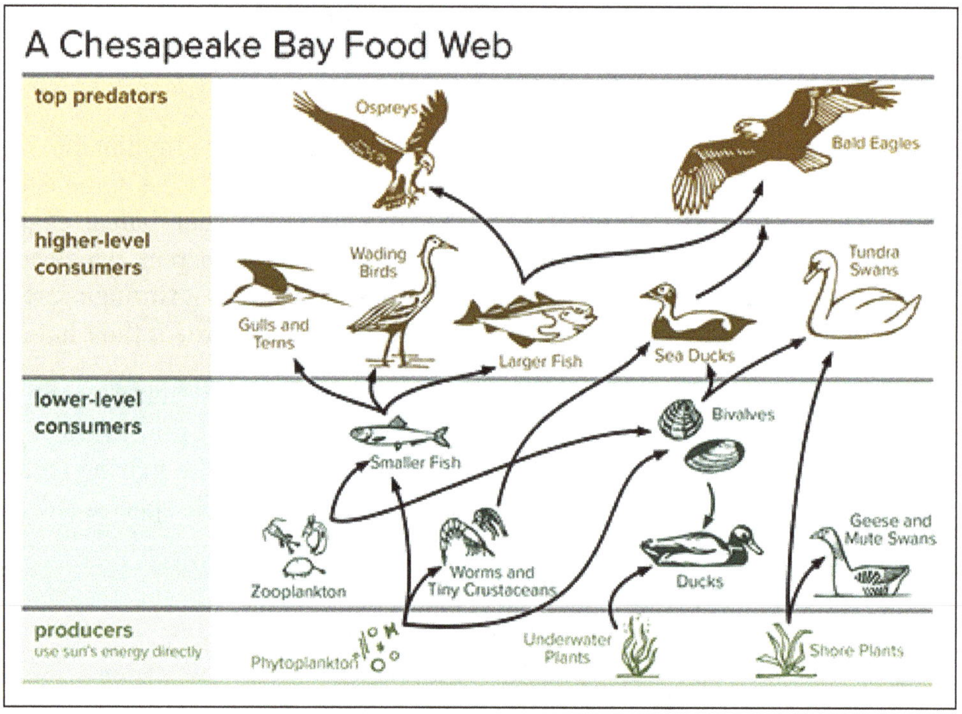

Waterbird food web in Chesapeake Bay.

Marine birds are adapted to life within the marine environment. They are often called *seabirds*. While marine birds vary greatly in lifestyle, behaviour and physiology, they often exhibit striking convergent evolution, as the same environmental problems and feeding niches have resulted in similar adaptations. Examples include albatross, penguins, gannets, and auks.

In general, marine birds live longer, breed later and have fewer young than terrestrial birds do, but they invest a great deal of time in their young. Most species nest in colonies, which can vary in size from a few dozen birds to millions. Many species are famous for undertaking long annual migrations, crossing the equator or circumnavigating the Earth in some cases. They feed both at the ocean's surface and below it, and even feed on each other. Marine birds can be highly pelagic, coastal, or in some cases spend a part of the year away from the sea entirely. Some marine birds plummet from heights, plunging through the water leaving vapour-like trails, similar to that of fighter planes. Gannets plunge into the water at up to 100 kilometres per hour (60 mph). They have air sacs under their skin in their face and chest which act like bubble-wrap, cushioning the impact with the water.

European herring gull attack herring schools from above.

Gentoo penguin swimming underwater.

The first marine birds evolved in the Cretaceous period, and modern marine bird families emerged in the Paleogene.

Mammals

Mammals are characterised by the presence of mammary glands which in females produce milk for feeding (nursing) their young. There are about 130 living and recently extinct marine mammal species such as seals, dolphins, whales, manatees, sea otters and polar bears. They do not represent a distinct taxon or systematic grouping, but are instead unified by their reliance on the marine environment for feeding. Both cetaceans and sirenians are fully aquatic and therefore are obligate water dwellers. Seals and sea-lions are semiaquatic; they spend the majority of their time in the water, but need to return to land for important activities such as mating, breeding and molting. In contrast, both otters and the polar bear are much less adapted to aquatic living. Their diet varies considerably as well: some may eat zooplankton; others may eat fish, squid, shellfish, and sea-grass; and a few may eat other mammals.

Sea otter, a classic keystone species which controls sea urchin numbers.

In a process of convergent evolution, marine mammals, especially cetaceans redeveloped their body plan to parallel the streamlined fusiform body plan of pelagic fish. Front legs became flippers and back legs disappeared, a dorsal fin reappeared and the tail morphed into a powerful horizontal fluke. This body plan is an adaptation to being an active predator in a high drag environment. A parallel convergence occurred with the now extinct marine reptile ichthyosaur.

Bottlenose dolphin, highest encephalization of any animal after humans.

Primary Producers

Primary producers are the autotroph organisms that make their own food instead of eating other organisms. This means primary producers become the starting point in the food chain for heterotroph organisms that do eat other organisms. Some marine primary producers are specialised bacteria and archaea which are chemotrophs, making their own food by gathering around hydrothermal vents and cold seeps and using chemosynthesis. However most marine primary production comes from organisms which use photosynthesis on the carbon dioxide dissolved in the water. This process uses energy from sunlight to convert water and carbon dioxide into sugars that can be used both as a source of chemical energy and of organic molecules that are used in the structural components of cells. Marine primary producers are important because they underpin almost all marine animal life by generating most of the oxygen and food that provide other organisms with the chemical energy they need to exist.

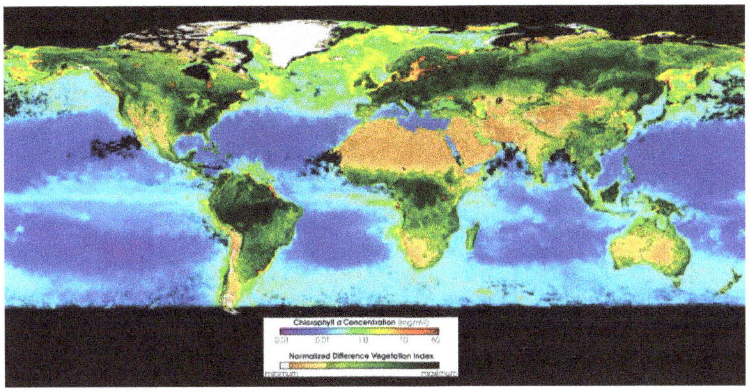

Composite image showing the global distribution of photosynthesis, including both oceanic phytoplankton and terrestrial vegetation. Dark red and blue-green indicate regions of high photosynthetic activity in the ocean and on land, respectively.

The principal marine primary producers are cyanobacteria, algae and marine plants. The oxygen released as a by-product of photosynthesis is needed by nearly all living things to carry out cellular respiration. In addition, primary producers are influential in the global carbon and water cycles. They stabilize coastal areas and can provide habitats for marine animals. The term division has been traditionally used instead of phylum when discussing primary producers, but the International Code of Nomenclature for algae, fungi, and plants now accepts both terms as equivalents.

Cyanobacteria

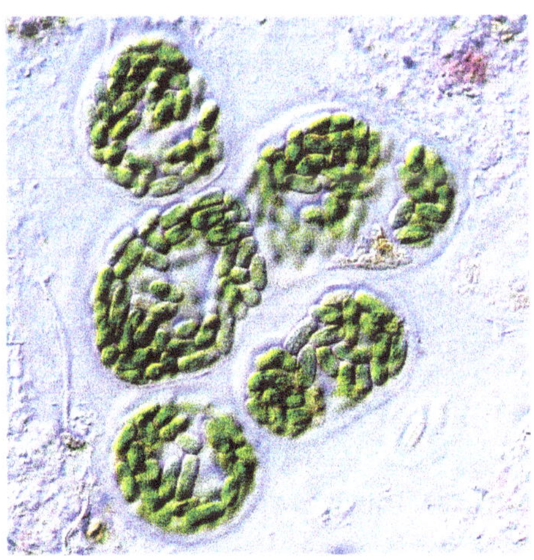

Cyanobacteria from a microbial mat. Cyanobacteria were the first organisms to release oxygen via photosynthesis.

Cyanobacteria are a phylum (division) of bacteria which range from unicellular to filamentous and include colonial species. They are found almost everywhere on earth: in damp soil, in both freshwater and marine environments, and even on Antarctic rocks. In particular, some species occur as drifting cells floating in the ocean, and as such were amongst the first of the phytoplankton.

The first primary producers that used photosynthesis were oceanic cyanobacteria about 2.3 billion years ago. The release of molecular oxygen by cyanobacteria as a by-product of photosynthesis induced global changes in the Earth's environment. Because oxygen was toxic to most life on Earth at the time, this led to the near-extinction of oxygen-intolerant organisms, a dramatic change which redirected the evolution of the major animal and plant species.

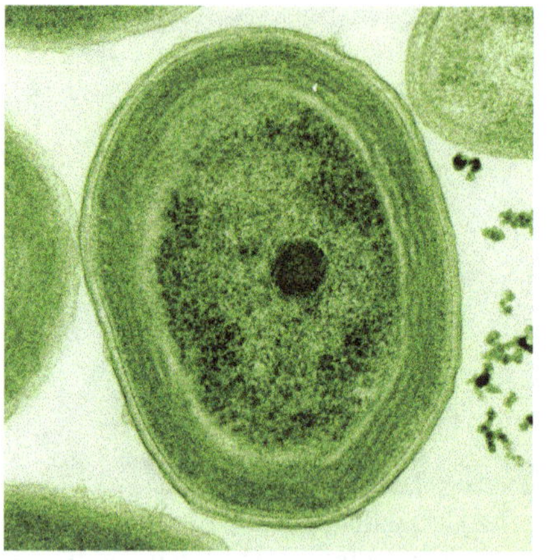

Prochlorococcus marinus

The tiny marine cyanobacterium *Prochlorococcus*, discovered in 1986, forms today part of the base of the ocean food chain and accounts for more than half the photosynthesis of the open ocean and an estimated 20% of the oxygen in the Earth's atmosphere. It is possibly the most plentiful genus on Earth: a single millilitre of surface seawater may contain 100,000 cells or more.

Originally, biologists thought cyanobacteria was algae, and referred to it as "blue-green algae". The more recent view is that cyanobacteria is a bacteria, and hence is not even in the same Kingdom as algae. Most authorities exclude all prokaryotes, and hence cyanobacteria from the definition of algae.

Algae

Algae is an informal term for a widespread and diverse group of photosynthetic protists which are not necessarily closely related and are thus polyphyletic. Marine algae can be divided into six groups:

- Green algae, an informal group containing about 8,000 recognised species. many species live most of their lives as single cells or are filamentous, while others form colonies made up from long chains of cells, or are highly differentiated macroscopic seaweeds.
- Red algae, a (disputed) phylum containing about 7,000 recognised species, mostly multicellular and including many notable seaweeds.
- Brown algae, a class containing about 2,000 recognised species, mostly multicellular and including many seaweeds, including kelp.
- Diatoms, a (disputed) phylum containing about 100,000 recognised species of mainly unicellular algae. Diatoms generate about 20 percent of the oxygen produced on the planet each year, take in over 6.7 billion metric tons of silicon each year from the waters in which they live, and contribute nearly half of the organic material found in the oceans. The shells (frustules) of dead diatoms can reach as much as half a mile deep on the ocean floor.
- Dinoflagellates, a phylum of unicellular flagellates with about 2,000 marine species. Many dinoflagellates are known to be photosynthetic, but a large fraction of these are in fact mixotrophic, combining photosynthesis with ingestion of prey (phagotrophy). Some species

are endosymbionts of marine animals and play an important part in the biology of coral reefs. Others predate other protozoa, and a few forms are parasitic.

- Euglenophytes, a phylum of unicellular flagellates with only a few marine members.

Diatoms	
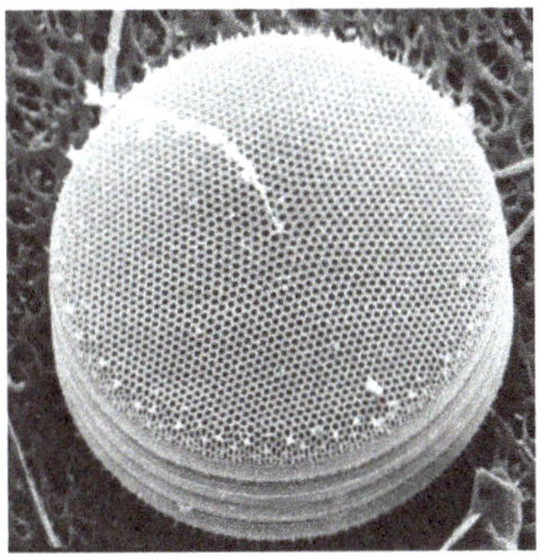	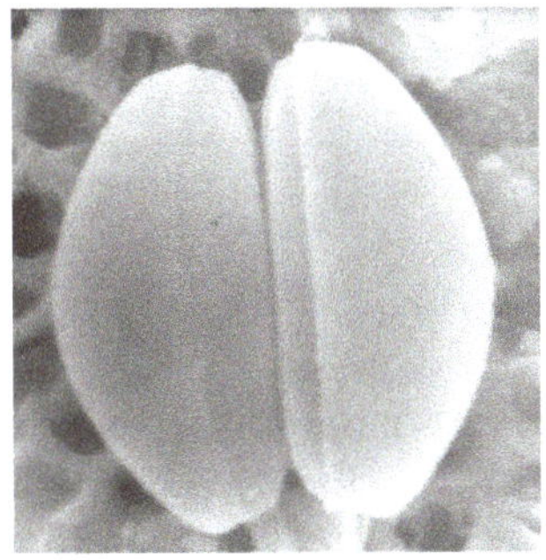
Centric	Pennate
Diatoms have a silica shell (frustule) with radial (centric) or bilateral (pennate) symmetry.	

Unlike higher plants, algae lack roots, stems, or leaves. They can be classified by size as *microalgae* or *macroalgae.*

Microalgae are the microscopic types of algae, not visible to the naked eye. They are mostly unicellular species which exist as individuals or in chains or groups, though some are multicellular. They are very diverse. It has been estimated there are 200,000-800,000 species of which about 50,000 species have been described. Depending on the species, their sizes range from a few micrometers (μm) to a few hundred micrometers. They are specially adapted to an environment dominated by viscous forces.

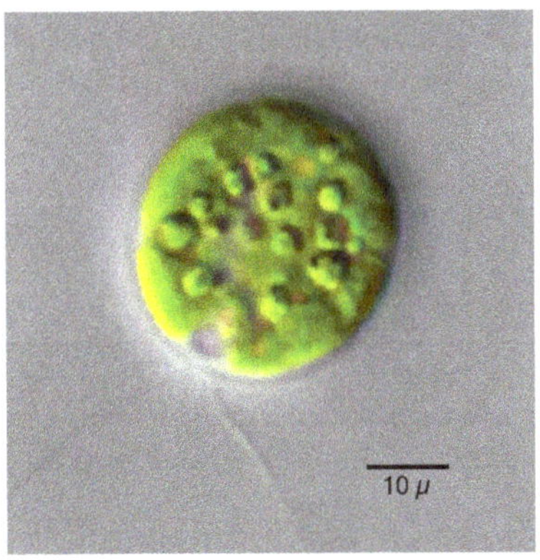

Chlamydomonas globosa, a unicellular green alga with two flagella just visible at bottom left.

Macroalgae are the larger, multicellular and more visible types of algae, commonly called seaweeds. Seaweeds usually grow in shallow coastal waters where they are anchored to the seafloor by a holdfast. Seaweed that becomes adrift can wash up on beaches. Kelp is a large brown seaweed that forms large underwater forests covering about 25% of the world coastlines. They are among the most productive and dynamic ecosystems on Earth. Some Sargassum seaweeds are planktonic (free-floating). Like microalgae, macroalgae (seaweeds) are technically marine protists since they are not true plants.

A seaweed is a macroscopic form of red or brown or green algae.

Unicellular Macroalgae

Unicellular organisms are usually microscopic, less than one tenth of a millimeter long. There are exceptions. Mermaid's wineglass, a genus of subtropical green algae, is single-celled but remarkably large and complex in form with a single large nucleus, making it a model organism for studying cell biology. Another single celled algae, *Caulerpa taxifolia*, has the appearance of a vascular plant including "leaves" arranged neatly up stalks like a fern. Selective breeding in aquariums to produce hardier strains resulted in an accidental release into the Mediterranean where it has become an invasive species known colloquially as *killer algae*.

Marine Plants

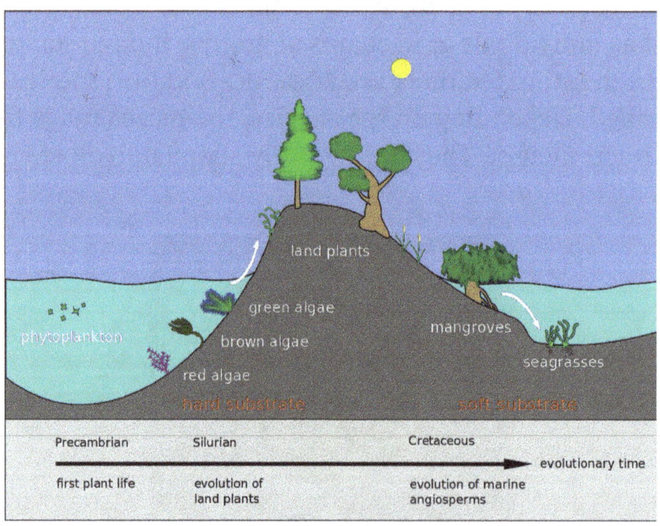

Evolution of mangroves and seagrasses.

Back in the Silurian, some phytoplankton evolved into red, brown and green algae. These algae then invaded the land and started evolving into the land plants we know today. Later, in the Cretaceous, some of these land plants returned to the sea as mangroves and seagrasses.

Marine plants can be found in intertidal zones and shallow waters, such as seagrasses like eelgrass and turtle grass, *Thalassia*. These plants have adapted to the high salinity of the ocean environment. Plant life can also flourish in the brackish waters of estuaries, where mangroves or cordgrass or beach grass beach grass might grow.

The total world area of mangrove forests was estimated in 2010 as 134,257 square kilometres

(51,837 sq mi) (based on satellite data). The total world area of seagrass meadows is more difficult to determine, but was conservatively estimated in 2003 as 177,000 square kilometres (68,000 sq mi).

Mangroves

Mangroves and seagrasses provide important nursery habitats for marine life, acting as hiding and foraging places for larval and juvenile forms of larger fish and invertebrates.

Plankton and Trophic Interactions

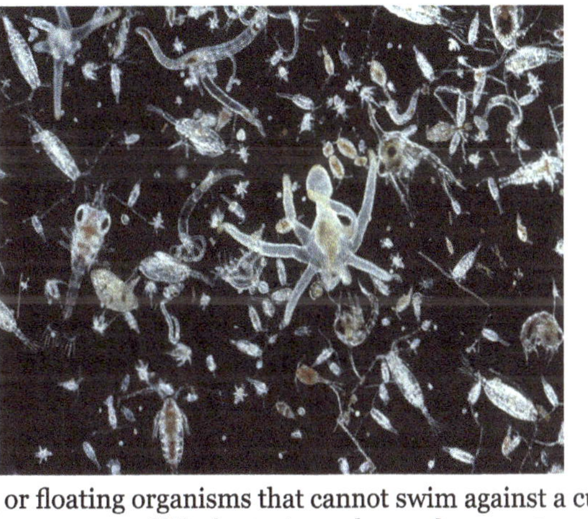

Plankton are drifting or floating organisms that cannot swim against a current, and include organisms from most areas of life: bacteria, archaea, algae, protozoa and animals.

Plankton are a diverse group of organisms that live in the water column of large bodies of water but cannot swim against a current. As a result, they wander or drift with the currents. Plankton are defined by their ecological niche, not by any phylogenetic or taxonomic classification. They are a crucial source of food for many marine animals, from forage fish to whales. Plankton can be divided into a plant-like component and an animal component.

Phytoplankton

Phytoplankton are the plant-like components of the plankton community. They are autotrophic (self-feeding), meaning they generate their own food and do not need to consume other organisms.

Phytoplankton consists mainly of microscopic photosynthetic eukaryotes which inhabit the upper sunlit layer in all oceans. They need sunlight so they can photosynthesize. Most phytoplankton are single-celled algae, but other phytoplankton are bacteria and some are protists. Phytoplankton can be categorised into cyanobacteria, various types of algae (red, green, brown, and yellow-green), diatoms, dinoflagellates, euglenoids, coccolithophorids, cryptomonads, chrysophytes, chlorophytes, prasinophytes, and silicoflagellates. They form the base of the primary production that drives the ocean food web, and account for half of the current global primary production, more than the terrestrial forests.

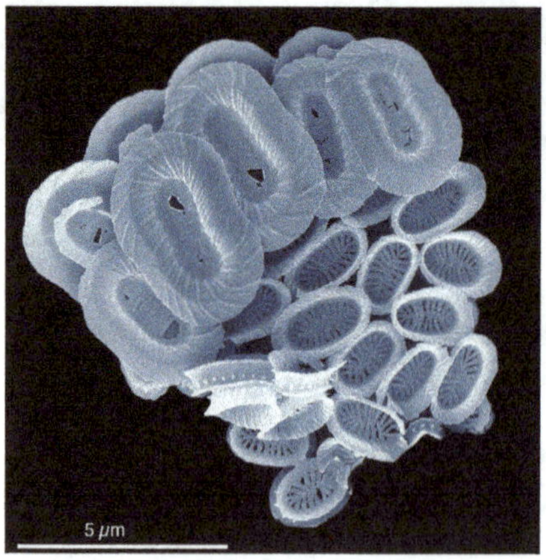

Widely distributed coccolithophores.

Zooplankton

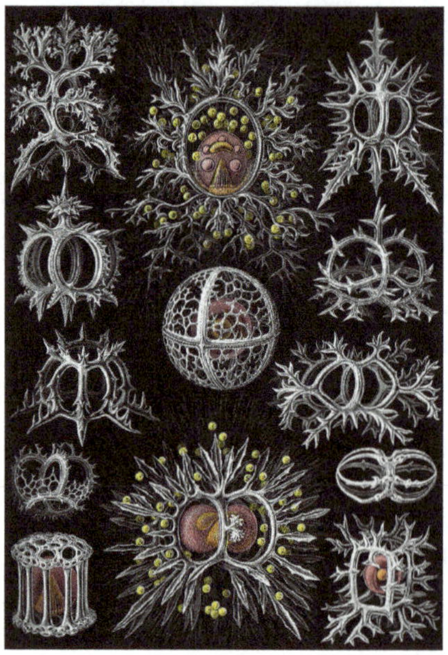

Radiolarian protists.

Zooplankton are the animal component of the planktonic community. They are heterotrophic (other-feeding), meaning they cannot produce their own own food and must consume instead other plants or animals as food. In particular, this means they eat phytoplankton.

Zooplankton are generally larger than phytoplankton, mostly still microscopic but some can be seen with the naked eye. Many protozoans (single-celled protists that prey on other microscopic life) are zooplankton, including zooflagellates, foraminiferans, radiolarians and some dinoflagellates. Other dinoflagellates are mixotrophic and could also be classified as phytoplankton; the distinction between plants and animals often breaks down in very small organisms. Other zooplankton include pelagic cnidarians, ctenophores, molluscs, arthropods and tunicates, as well as planktonic arrow worms and bristle worms.

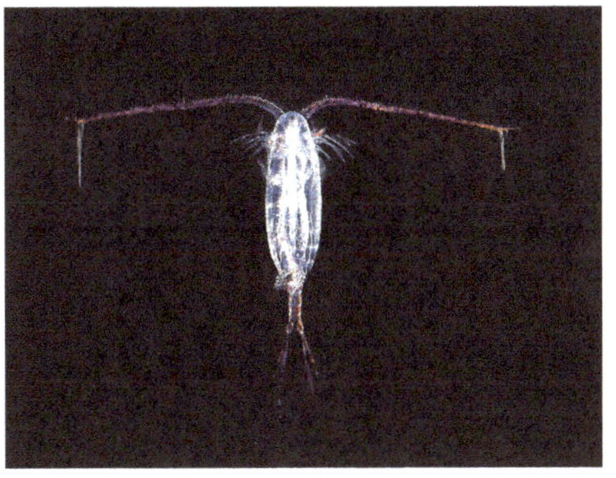

Copepod eat phytoplankton. This one is carrying eggs.

Larger zooplankton can be predatory on smaller zooplankton.

Moon jellyfish

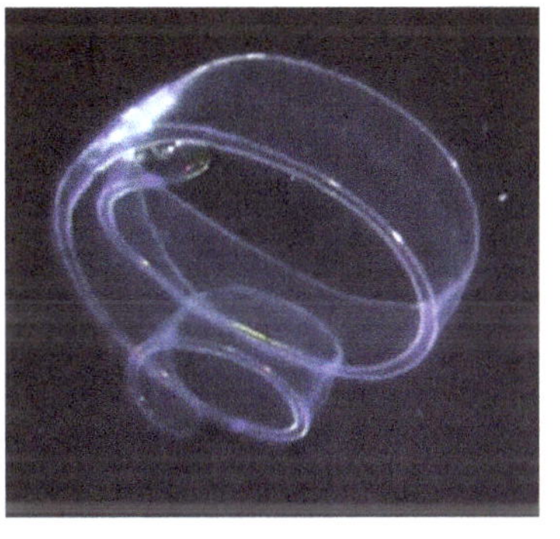

Venus girdle, a ctenophore

Many marine animals begin life as zooplankton in the form of eggs or larvae, before they develop into adults. These are meroplanktic, that is, they are planktonic for only part of their life.

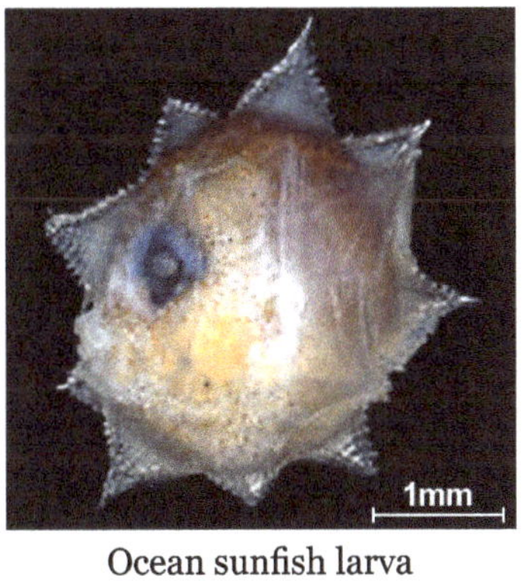

Ocean sunfish larva

Mixotrophic Plankton

Dinoflagellates are often mixotrophic or live in symbiosis with other organisms. Some dinoflagellates are bioluminescent. At night, ocean water can light up internally and sparkle with blue light because of these dinoflagellates. Bioluminescent dinoflagellates possess scintillons, individual cytoplasmic bodies which contain dinoflagellate luciferase, the main enzyme involved in the luminescence. The luminescence, sometimes called the phosphorescence of the sea, occurs as brief (0.1 sec) blue flashes or sparks when individual scintillons are stimulated, usually by mechanical disturbances from, for example, a boat or a swimmer or surf.

A surf wave at night sparkles with blue light due to the presence of a bioluminescent dinoflagellate, such as *Lingulodinium polyedrum*.

Marine Food Web

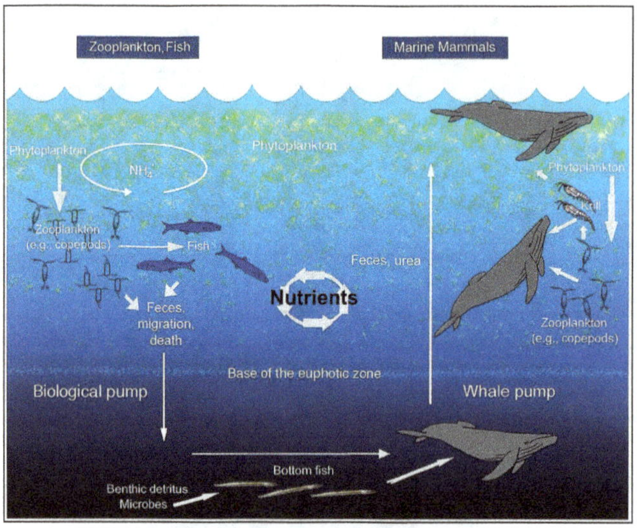

Whale pump nutrient cycle.

In 2010 researchers found whales carry nutrients from the depths of the ocean back to the surface using a process they called the whale pump. Whales feed at deeper levels in the ocean where krill is found, but return regularly to the surface to breathe. There whales defecate a liquid rich in nitrogen and iron. Instead of sinking, the liquid stays at the surface where phytoplankton consume it. In the Gulf of Maine the whale pump provides more nitrogen than the rivers.

If phytoplankton dies before it is eaten, it descends through the euphotic zone and settles into the depths of sea. In this way, phytoplankton sequester about 2 billion tons of carbon dioxide into the ocean each year, causing the ocean to become a sink of carbon dioxide holding about 90% of all sequestered carbon.

Zooplankton make up most of the marine animal biomass, and as primary consumers are the crucial link between primary producers (mainly phytoplankton) and the rest of the marine food web (secondary consumers).

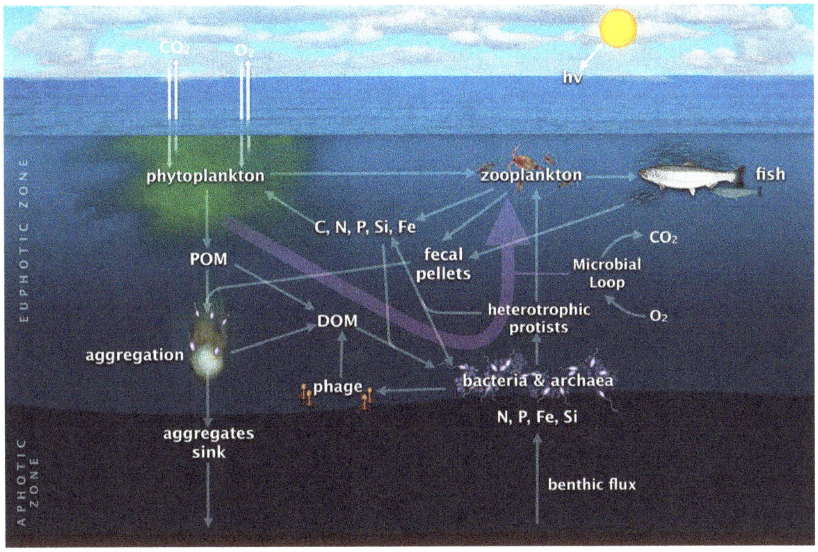

Marine food web.

Sediments and Biogenic Ooze

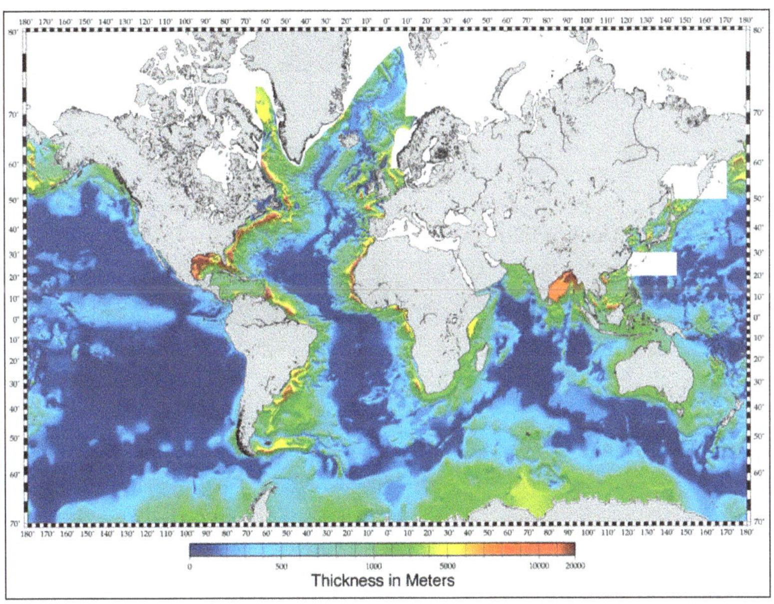

Thickness of marine sediments.

Sediments at the bottom of the ocean have two main origins, terrigenous and biogenous. Terrigenous sediments account for about 45% of the total marine sediment, and originate in the erosion of rocks on land, transported by rivers and land runoff, windborne dust, volcanoes, or grinding by glaciers.

Biogenous sediments account for the other 55% of the total sediment, and originate in the skeletal remains of marine protists, single-celled plankton and benthos organisms. Much smaller amounts of precipitated minerals and meteoric dust can also be present. *Ooze*, in the context of a marine sediment, does not refer to the consistency of the sediment but to its biological origin. The term ooze was originally used by John Murray, the "father of modern oceanography", who proposed the term *radiolarian ooze* for the silica deposits of radiolarian shells brought to the surface during the Challenger Expedition. These days, biogenic ooze refers to pelagic sediments with at least 30 percent made up of the skeletal remains of marine organisms.

Main types of biogenic ooze					
Type	Mineral forms	Protist responsible		Name of skeleton	Description
Siliceous ooze	SiO_2 quartz glass opal-chert	Diatoms		Frustule	Individual diatoms range in size from 0.002 to 0.2 mm
		Radiolarians	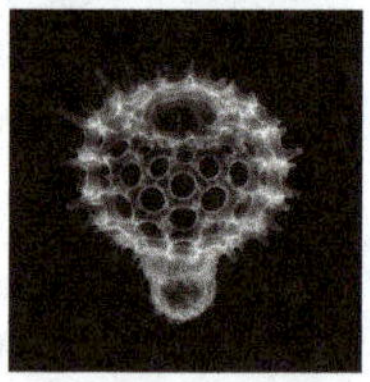	Skeleton	Radiolarians are protozoa with diameters between 0.1 and 0.2 mm that produce intricate mineral skeletons, usually made of silica
Calcareous ooze	$CaCO_3$ calcite aragonite limestone chalk	Foraminiferans	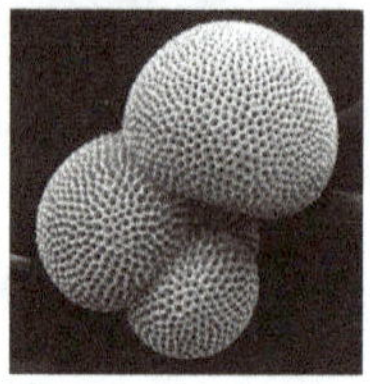	Test	There are about 10,000 living species of foraminiferans, usually under 1 mm in size
		Coccolithophores	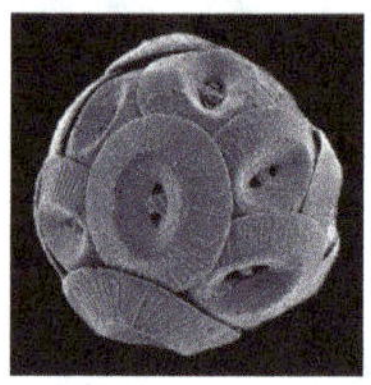	Coccolith	Coccolithophores are spherical cells usually less than 0.1 mm across, enclosed by calcareous plates called coccoliths. Coccoliths are important microfossils. They are the main constituent of chalk deposits such as the white cliffs of Dover

Biogeochemical Cycles

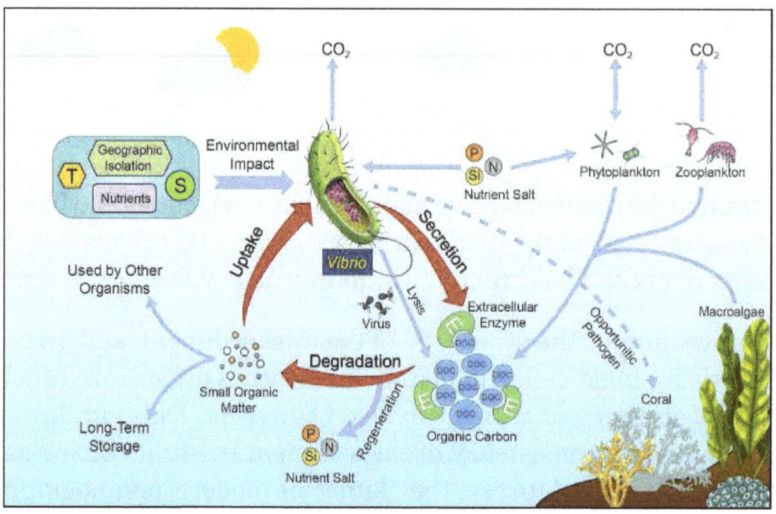

Marine carbon cycle

In 2000 a team of microbiologists led by Edward DeLong made a crucial discovery in the understanding of the marine carbon and energy cycles. They discovered a gene in several species of bacteria responsible for production of the protein rhodopsin, previously unheard of in the domain Bacteria. These proteins found in the cell membranes are capable of converting light energy to biochemical energy due to a change in configuration of the rhodopsin molecule as sunlight strikes it, causing the pumping of a proton from inside out and a subsequent inflow that generates the energy.

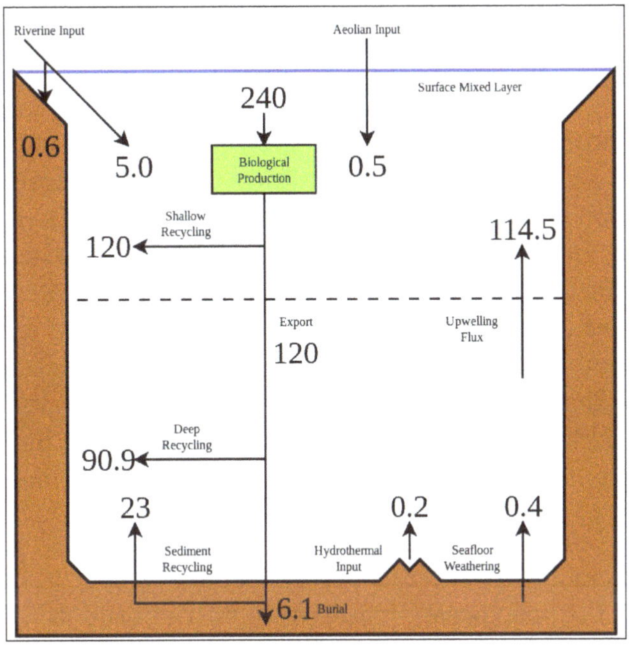

Marine silicon cycle

Anthropogenic Impacts

Fishing down the foodweb: Overfishing of high trophic fish like tuna can result in them being replaced by low trophic organisms, like jellyfish.

Overfishing is occurring in one third of world fish stocks, according to a 2018 report by the Food and Agriculture Organization of the United Nations. In addition, industry observers believe illegal, unreported and unregulated fishing occurs in most fisheries, and accounts for up to 30% of total catches in some important fisheries. In a phenomenon called fishing down the foodweb, the mean trophic level of world fisheries has declined because of overfishing high trophic level fish.

Ocean acidification is the increasing acidification of the oceans, caused by the uptake of carbon dioxide from the atmosphere. The rise in atmospheric carbon dioxide due to the burning of fossil fuels leads to more carbon dioxide dissolving in the ocean. When carbon dioxide dissolves in water it forms hydrogen and carbonate ions. This in turn increases the acidity of the ocean and makes survival increasingly harder for shellfish and other marine organisms that depend on calcium carbonate to form their shells. Increasing acidity is also thought to have a range of potentially harmful consequences for marine organisms, such as depressing metabolic rates and immune responses in some organisms, and causing coral bleaching. Ocean acidification has been compared to anthropogenic climate change and called the "evil twin of global warming" and "the other CO_2 problem".

Marine pollution results from the entry into the ocean of industrial, agricultural, and residential wastes. Pathways for this pollution include agricultural runoff into rivers and wind-blown debris and dust. Nutrient pollution is a primary cause of eutrophication of surface waters, in which excess nutrients, usually nitrates or phosphates, stimulate algae growth. Toxic chemicals can adhere to tiny particles which are then taken up by plankton and benthic animals, most of which are either deposit feeders or filter feeders. In this way, toxins are concentrated upward within ocean food chains. Many particles combine chemically in a manner which depletes oxygen, causing estuaries to become anoxic. Pesticides and toxic metals are similarly incorporated into marine food webs, harming the biological health of marine life. Many animal feeds have a high fish meal or fish hydrolysate content. In this way, marine toxins are transferred back to farmed land animals, and then to humans.

Phytoplankton concentrations have increased over the last century in coastal waters, and more recently have declined in the open ocean. Increases in nutrient runoff from land may explain the increases in coastal phytoplankton, while warming surface temperatures in the open ocean may have strengthened stratification in the water column, reducing the flow of nutrients from the deep that open ocean phytoplankton find useful.

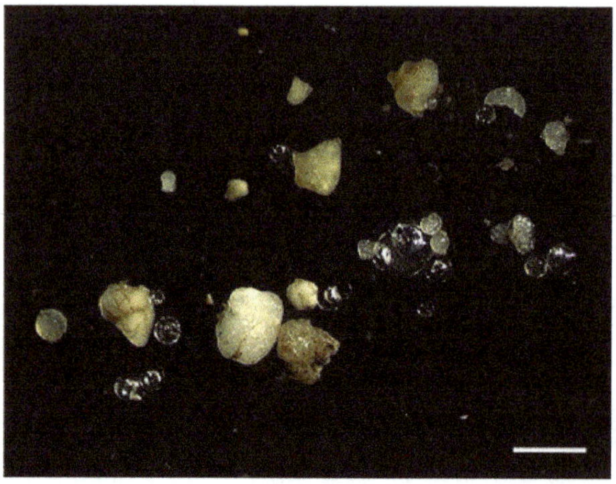

Microplastics among sand and glass spheres in sediment from the Rhine. The white bar represents 1 mm.

Estimates suggest something like 9 million tonnes of plastic is added to the ocean every year. This plastic may need 450 years or more to biograde. Once in the ocean, plastics are shredded by marine amphipods into microplastics. There are now beaches where 15 percent of the sand are grains of microplastic. In the oceans themselves, microplastics float in surface waters amongst the plankton where they are ingested by plankton eaters.

Habitat loss is occurring in seagrass meadows, mangrove forests, coral reefs and kelp forests, all of which are in global decline due to human disturbances. Seagrasses have lost 30,000 km^2 (12,000 sq mi) during recent decades, one fifth of the world's mangrove forests have been lost since 1980, and another fifth of coral reefs have been lost. The most pressing threat to kelp forests may be the overfishing of coastal ecosystems, which by removing higher trophic levels facilitates their shift to depauperate urchin barrens.

Shifting baselines arise in research on marine ecosystems because changes must be measured against some previous reference point (baseline), which in turn may represent significant changes from an even earlier state of the ecosystem. For example, radically depleted fisheries have been evaluated by researchers who used the state of the fishery at the start of their careers as the baseline, rather than the fishery in its unexploited or untouched state. Areas that swarmed with a particular species hundreds of years ago may have experienced long term decline, but it is the level a few decades previously that is used as the reference point for current populations. In this way large declines in ecosystems or species over long periods of time were, and are, masked. There is a loss of perception of change that occurs when each generation redefines what is natural or untouched.

Biodiversity and Extinction Events

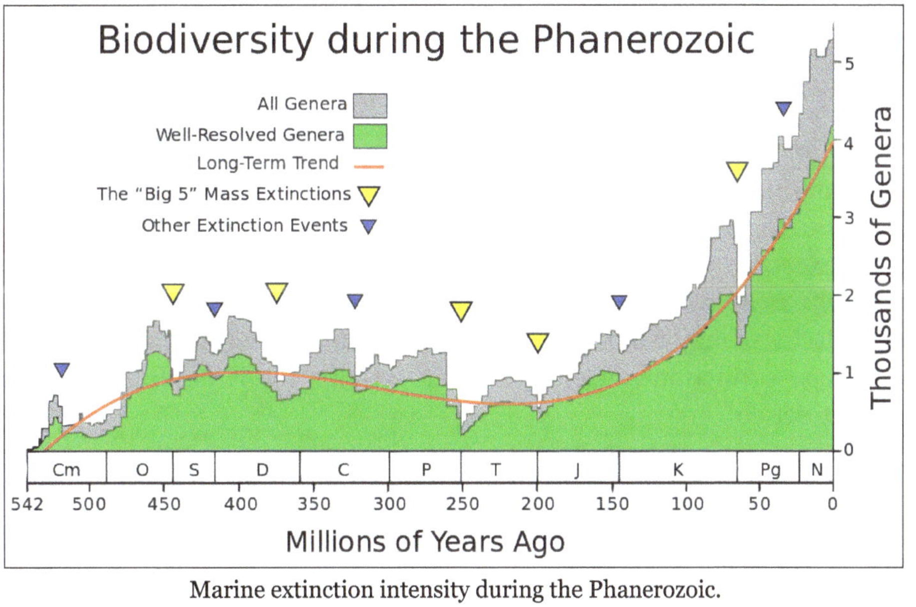

Marine extinction intensity during the Phanerozoic.

Biodiversity is the result of over three billion years of evolution. Until approximately 600 million years ago, all life consisted of archaea, bacteria, protozoans and similar single-celled organisms. The history of biodiversity during the Phanerozoic (the last 540 million years), starts with rapid growth during the Cambrian explosion – a period during which nearly every phylum of multicellular organisms first appeared. Over the next 400 million years or so, invertebrate diversity showed little overall trend and vertebrate diversity shows an overall exponential trend.

However, more than 99 percent of all species that ever lived on Earth, amounting to over five billion species, are estimated to be extinct. These extinctions occur at an uneven rate. The dramatic rise in diversity has been marked by periodic, massive losses of diversity classified as mass extinction events. Mass extinction events occur when life undergoes precipitous global declines.

Most diversity and biomass on earth is found among the microorganisms, which are difficult to measure. Recorded extinction events are therefore based on the more easily observed changes in the diversity and abundance of larger multicellular organisms, rather than the total diversity and abundance of life. Marine fossils are mostly used to measure extinction rates because of their superior fossil record and stratigraphic range compared to land organisms.

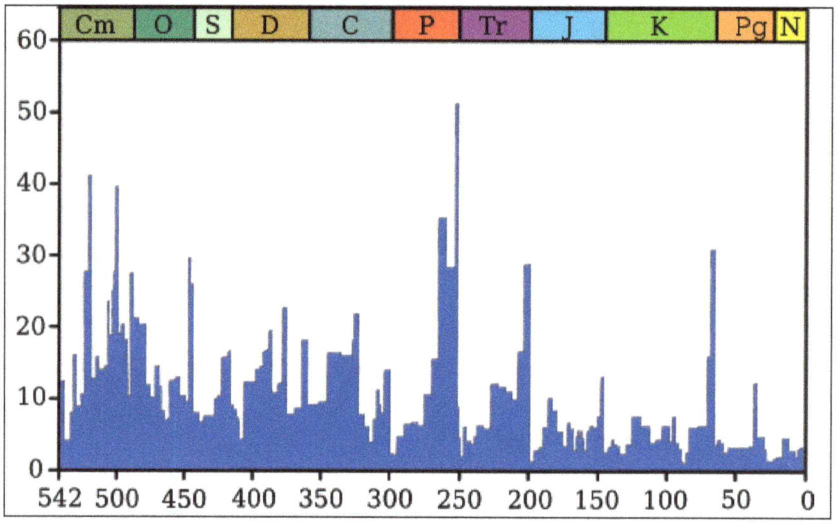

Apparent extinction intensity, i.e. the fraction of genera going extinct at any given time as reconstructed from the fossil record (excluding the current Holocene extinction event).

Based on the fossil record, the background rate of extinctions on Earth is about two to five taxonomic families of marine animals every million years. The Great Oxygenation Event was perhaps the first major extinction event. Since the Cambrian explosion five further major mass extinctions have significantly exceeded the background extinction rate. The worst was the Permian-Triassic extinction event, 251 million years ago. Vertebrates took 30 million years to recover from this event. In addition to these major mass extinctions there are numerous minor ones, as well as the current ongoing mass-extinction caused by human activity, the Holocene extinction sometimes called the "sixth extinction".

Marine Biology

Marine biology is the study of marine ecosystems, marine organisms, and human interactions with these environments and species. As a marine biologist, you may study a wide range of topics from the behavior, physiology, and ecology of specific groups of organisms to the interactions between human activity and aquatic environments.

The organisms that are studied by marine biologists vary greatly and include everything from the biggest whales to the smallest plankton. Marine biologists also study abiotic factors within the environment, such as the concentration of nutrients within the ocean. These types of studies help scientists understand the ocean in greater detail.

Since the Earth consists of approximately 71% water. This means that the majority of the Earth is covered in oceans. Marine biologists study every aspect of oceans, as they contribute many things humans need to survive. This could be the oxygen produced by algae in the ocean, to the fish harvested by billions of people every year for food.

Branches of Marine Biology

By type of organism:

- Herpetology: The study of amphibians and reptiles, many of which live in or around the ocean.
- Ichthyology: The study of fish.
- Invertebrate zoology: The study of invertebrate organisms (organisms without a backbone). Many invertebrates are aquatic including sponges, crustaceans, and mollusks.
- Marine mammology: The study of cetaceans, which includes whales, dolphins, and porpoises.
- Phycology: The study of algae.

Other fields that are not necessarily organism-specific:

- Environmental Studies: Marine biology is closely intertwined with the study of the environment. Marine biologists study how aquatic organisms are affected by rising ocean temperatures, water pollution, and ocean acidification.
- Fisheries and aquaculture: These fields involve the use of fish as a food source for humans. Fisheries are where fish are raised/farmed to be eaten. Some marine biologists work with fisheries and conduct research on overfishing, the sustainability of fish as a food source, and pollution levels. Aquaculture refers to research on farming specific species of fishes in order to learn how to best cultivate them as a food source.
- Marine biotechnology: This field involves biomedical applications of marine life. For example, scientists study how compounds from marine plants, animals, and microbes may be used in pharmaceuticals. The drug Ara-C, used to treat leukemia and non-Hodgkin's lymphoma, and the drug Ara-A, used to treat herpes, are both derived from compounds found in a sea sponge.

Threats to Marine Ecosystems

Ocean Dumping

Dumping of waste materials from industries, ships and sewage plants into oceans has polluted the marine ecosystem to a great extent. For years the oceans have been targeted as a dumping site of sewage, chemicals, industrial waste, garbage, and other debris from the land. According to reports, only mining companies across the world dump 220 million tonnes hazardous waste directly into our water bodies every year. Similarly, it is important to note that around two-thirds of the marine lives in the world have been under threat from the chemicals we use every day, including household cleaners. Since we depend on the marine ecosystem extensively, the adverse effects of ocean dumping have not only been felt by marine life but by humans as it poses health risks.

Land Runoff

One of the major sources of ocean pollution is the waste comes from the nonpoint source, which happens as a result of runoff. Surface runoff from both agricultural land and areas carry soil and particles mixed with carbon, phosphorus, nitrogen and minerals, posing threat to the marine life in alarming scale. Crossing streams and rivers, the water filled with these toxic chemicals land in the ocean, resulting in harmful algal blooms. This kind of water pollution threatens the species of fish, turtles, and shrimp etc. and also humans through the food chain.

Dredging

In this world of ever expanding industrial activities, dredging is an important activity that enhances marine transportation and other related activities. However, dredging has been a major cause of disturbance in the marine ecosystem for many years. As dredging is to remove the deposits submerged underwater, the activity alters the pre-disposed composition of soil, leading to the destruction habitat of creatures and organisms. Similarly, dredging of contaminated materials will result in the regrouping of harmful particles and contaminate large of areas of water bodies. Though steps have been taken to mitigate the effects of dredging on the marine environment, several cases involving the destruction of underwater lives are still extensively heard about.

NOx and SOx

Nitrogen oxides (NOx) and Sulfur oxides (SOx) – the two main pollutants found in shipping emissions- has badly affected both marine environment and ozone layer in a number of ways. Both NOx and SOx are combustion products that are emitted into the environment in the form of smoke. It is estimated that in 2005, the water bodies around Europe witnessed 1.7 million tonnes of sulphur dioxide (SO_2) emissions and 2.8 million tonnes nitrogen dioxide emissions from international shipping. And, according to the recent studies, these type of air pollution from shipping accounts around for 50,000 premature deaths per year in Europe. However, strict rules have been formed to reduce their levels in ship's emissions. With IMO revising its standards on the sulphur content of marine fuels, the vessels passing through a Sulphur Emission Control Area (SECA) are not permitted, since 2015, to use fuels with more than 0.1% of sulphur. Similarly, the sulphur limit applicable to all marine fuels used internationally will go from 3.5% to 0.5% since 2020.

Ocean Acidification

The issue of ocean acidification is quickly becoming a threat to both marine lives and humans. Ocean acidification is the continuing decrease of seawater pH caused by the absorption of carbon dioxide (CO_2) from the atmosphere. Ocean acidification has the power to greatly endanger the lives of marine organisms and also humans who depend on fish and fish products for their daily sustenance. Studies have shown that the decreased pH levels affect the behaviour of several marine species, putting them at life-threatening risks.

Sea Water Level Rising

Global warming is alarmingly increasing seawater levels, threatening the marine ecosystem. According to reports, the annual rate of seawater rise during the past two decades has been 0.13 inches a year, which is around twice the average speed of rising over the preceding 80 years. Thus, it is high time that we educate ourselves about the causes and effects of seawater level rising to save the marine environment and biodiversity. The rise in seawater level means more wetland flooding, destructive erosion and agricultural land contamination and more importantly a serious threat to the habitat of several plants, fishes and birds.

Ozone Depleting Substances

Ozone-depleting substances such as CFCs and Halon along with other pollutants from ships are destroying the ozone layer. Ozone Depleting Substances omitted by ships across the world include Methyl Chloroform, Methyl Bromide, Bromochlorodifluoromethane and Bromotrifluoromethane etc. These man-made gases are capable of destroying ozone and in effect, these gases causing harm to the marine environment in several ways.

Waste Pollution from Ships

As we know, tens and thousands of ships are responsible for more than 90 per cent of world trade. Apart from other pollutants such as oil and gas, the waste and garbage generated on board ships poses a significant threat to the marine ecosystem. Both solid and liquid waste in form of ballast water, grey water, food waste, dunnage and packing material, paper products and cleaning material and rags etc. pollutes the seawater and badly affects marine lives. The vessels used for various purposes-be it a container or cruise ship-contribute to this pollution in different levels.

Noise Pollution from Ships

It has been scientifically proved that the noise generated from shipping operations is harmful to marine organisms. Harmful effects of noise pollution on marine organisms include haemorrhages, changed diving pattern, migration to newer places, and damage to internal organs and an overall panic response to foreign sounds. Source of noise pollution from ships include everything from engine noise to the entertainments in cruise ships. The intensity of noise pollution is higher in marine environment since noise travels greater distances easily in water and at the same time, marine life is extremely sensitive to noise due to their heavy reliance on underwater sounds for basic life functions.

Oil Spills

The biggest cause of marine pollution – oil spills. The world has witnessed several oil spill disasters that have been one of the major concerns of pollution to the marine environment. Disasters such as the Exxon Valdez Oil Spill and Deepwater Horizon etc. have resulted in the extreme pollution of the marine ecosystem, killing thousands of marine species. The oil spilt destroys the insulating ability of several marine species and also the water repellency of bird's feathers, exposing these creatures to life-threatening risks.

Plastic Pollution

This environment degrading agent separately for the sole fact that it has and is the reason for several environmental problems both at sea and land. Those who have read or seen the Pacific garbage patch knows the extent of damage this substance is causing to the marine environment.

It is estimated that around 8 million tonnes of plastic waste enters our oceans every year, and by 2050, at this rate, we would witness more plastic than fish in the water bodies across the world. The ill effect of plastic pollution is wide-ranging. The plastic pollution has a direct effect on wildlife as it-plastic bags, fishing nets and other debris-chokes tens and thousands of seabirds and sea turtles every year. The ingestion micro plastics fish and other species also pose risk to their life as well as humans.

Salt Marsh

Salt marshes are defined as natural or semi-natural terrestrial halophytic ecosystems. They largely occur in the intertidal zone between land and the sea and are covered by salty or brackish water for at least some of the time. They replace mangroves in temperate and arctic regions. The dominant flora is composed of halophytic plant such as grasses, shrubs and herbs. The flora is rather species poor. The sediment consists of mud and sand. The salt marshes are normally associated with mud flats but also occur on sand flats. These mud flats are sometimes dominated by algae. They are regularly subjected to tidal movement, the height of which can vary from a few centimetres (in enclosed seas such as the Baltic Sea) to several metres in open sea areas such as the Atlantic. The drainage of this seawater is controlled by a meandering network of tidal channels. Through these channels sediments, detritus, dissolved nutrients, plankton and small fishes are flushed in and out the salt marshes.

Land of Saeftinghe - Belgium.

Distribution

Salt marshes are widely distributed in estuarine systems around the world. They also occur in deltas, rias and on open coasts. Although sediment is a prerequisite for their growth in height and width, salt marsh communities can occur in areas with limited or no sediment supply. Examples include sea water drenched cliffs and slopes on exposed coasts, at the head of sea lochs and rocky beaches. They have a range from the Arctic region over Europe, Africa, America, Asia to the coast of Australia.

The most extensive development of these salt marshes occurs in estuaries with a large tidal range, abundant fine-grained sediments, sheltered location where particles can settle out of the water column and moderate climate.

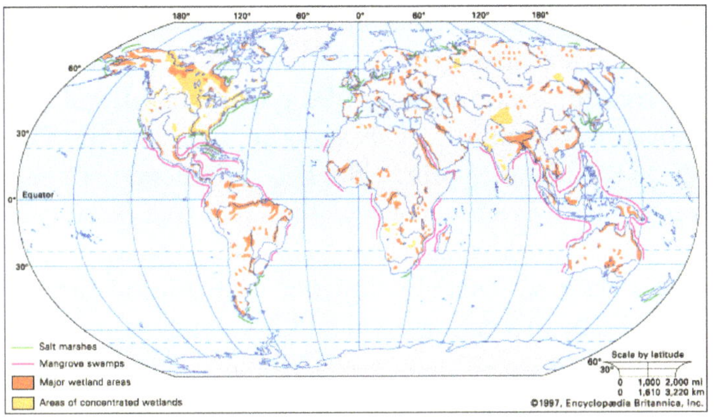

Distribution of salt marshes (green), wetlands (orange and yellow) and mangroves (pink).

Evolution

Salt marshes evolve over time from young marshes to old marshes. The natural young marshes in eastern USA consist for the largest part of low marsh cordgrass *Spartina alterniflora*. Nutrients are transported by tidal flooding through the tidal channels. This makes it possible for the grasses to grow thickly and abundantly what weakens the effect of waves and tidal currents so the depositional rate of mud increases. Erosion is reduced by the roots and rhizomes of the plants. At the time that the marsh surface builds up above the high water level, the high marsh plants invade, outcompete and replace the low marsh plants. When the quantity of the low and high marshes is equal, the ecosystem is in a mature stage of development. The continued deposition of mud converts most of the low marshes into high marshes. These are called the old marshes. Little water flows through the tidal channels and the marshes are elevated. At this time, streams and rivers deposit sand and mud on these high marshes and convert it into a dry land that is disconnected from ocean influences.

Requirements for Development

The requirements for development of salt marshes are:

- They need fine-grained sediments.
- There may be no strong waves or tidal currents.
- They need salty conditions to grow. They are halotolerant and have adaptations to these conditions.
- They need a temperate or cool temperature. Freezing temperatures can occur, but are not damaging the plants.
- They need a wide tidal range. This is important because it limits the erosion, makes deposition of sediments possible and causes a well-marked zonation.

Zonation

Based on the topography and characteristic plant assemblages, salt marshes are classified as low, medium and high marshes and is related to the number of tidal submergences per year. Below the low marshes are the tidal creeks. These creeks are channels that are flushed with salty or brackish water. The low marshes extend from the mean low water neap tide to the mean high water spring tide. An example from the eastern shores of the United States of America is dominated by the smooth cordgrass Spartina alterniflora. This grass spreads rapidly. The high marshes extend from the high water tide to the highest spring tide. It is only flooded during extremely high tides and during storms. It resembles more to terrestrial than to a true marine environment. The plant community is more diverse than the low marshes. It consists of Spartina patens, the saltwort grass Salicornia species and the spike grass Distichlis spicata. Distribution, density and activity of invertebrates are mainly controlled by protection, food and frequency of tidal flooding.

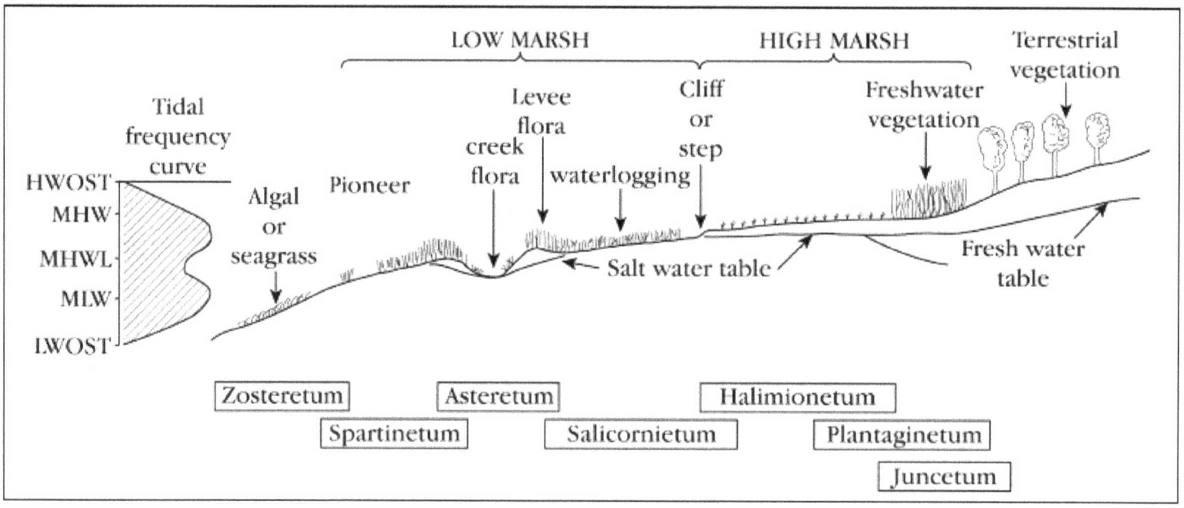

Zonation of salt marshes.

Succession

Succession is the serial development of different vegetation types at one place in time. It is a complex process and the factors determining.

The succession starts with unicellular algae such as diatoms sticking the sand together by production of mucus. This causes a brownish biofilm on the substrate. After this stage, filamentous algae optimize the fixation of sediments. These algae are mostly blue-green algae, Cyanophyta and Chlorophyta. Small gastropods can feed and develop in huge quantities on it. Locally, the filamentous alga Vaucheria can form banquettes or elevations. Brown algae are loosely associated with this stage.

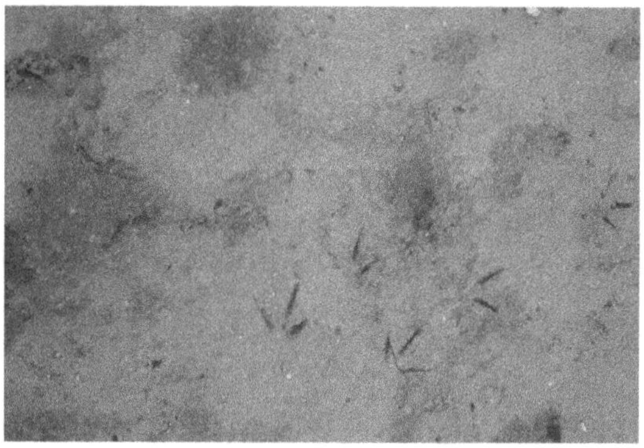

Biofilm of unicellular algae.

A next stage is the germination of species such as Salicornia. The seeds germinate after some desalination of the soil by rain. Between and around the glassworts takes extra sediment accumulation place. This results in elevating the substrate and makes it more stable. Other species such as Spartina maritima and Spartina anglica compete for the same place. S. maritima is an indigenous species of continental Europe and S. anglica is imported from the British Islands.

In Europe, other plants can occur. These herbs are Common Saltmarsh-grass (Puccinellia maritima), Sea Plantain (Plantago maritima), Sea Arrow-grass (Triglochin maritinum), Sea Aster (Aster tripolium), English Scurvy-grass (Cochlearia anglica), Common Sea-lavender (Limonium vulgare), Sea Purslane (Halimione portulacoides), the red alga Bostrychia scorpioides and Catenella caespitosa. Juncus maritimus is another species that can occur in salt marshes.

Salicornia perennis

Functioning and Adaptations

Salt marshes have several functions and adaptations to a life in an intertidal ecosystem. They need to conquer some problems to be resistant to the environment. Therefore, it is important to have some adaptations to survive. The first problem is that the plants are freshwater plants. This causes waterstress. For this reason, they have to take up water against the osmotic pressure. To overcome the negative osmotic pressure, they generate a negative hydrostatic pressure (by transpiration processes). They have thin, fleshy leaves and are sensitive to extra stress such as pollution. Anatomically, the plants are adapted through strong lignification, a well-developed epidermis and succulent leaves and stems. Evaporation can be limited by thin leaves with scale-like hairs. Physiologically, plants are adapted by accumulating salt in their tissues. In this way, normal osmosis is possible. Other plants have salt gland cells on the lower surface of the leaves and excrete the salt from its tissue.

A second problem is the anoxic environment. The underground in which they root is saturated with water. The tissue of the plants requires oxygen for respiration. Gas diffusion between gas particles can only supply this need in soils that are not waterlogged. Even when the water is saturated with oxygen, its concentration is too low and the diffusion in water is very slow. This is solved by aerenchyma, a tissue that can provide air to the submerged parts of the plants. Roots are superficial systems because of the anoxic sediments. These systems are composed of perennial thick roots with a corky layer and without root hairs. To fixate the substrate, shortliving thin and strongly branched roots with numerous root hairs are developed so they can take up nutrients. Salt marshes have a whole range of functions. It plays an important role as a sediment trap. In this way, it regulates the water quality and helps stabilizing the coastlines. Salt marshes function as filtering systems and retain sediments, excess nutrients, toxic chemicals and disease-causing organisms. They remove nitrates and phosphates from rivers and streams which receive waste water effluents. Another function is the recharge and discharge of groundwater attributing to the water supply. Salt marshes function also as important habitat, nursery grounds for larvae and small organisms, shelter, providing food and a nesting ground for wading birds and other organisms.

Fauna

Eurasian oystercatcher

Small mammals, small fishes, birds, insects, spiders, and marine invertebrates are found in salt marshes. Marine invertebrates include amphipods, isopods, anemones, shrimps, crabs, turtles, mollusks and gastropods. Fishes such as sticklebacks, silversides, eels and flounders are found in the waters of the salt marshes. The salt marshes are important breeding, feeding and overwintering grounds for waterfowl. These waterfowls consist of ducks, herons, Sharptailed Sparrows, Eurasian Oystercatchers, Reed Bunting. The benthic community consists of mollusks, polychaeta and oligochaeta. The coastal freshwater marshes are species poor. The organisms that you can find here are water boatmen, flies, mosquitos and snails. Further there are mollusks, ducks, geese, muskrats, raccoons, minks and other small mammals. For some species, a seasonal appearance can be seen. In Saudi Arabia, salt marshes are grazing places for wild dromedaries.

Water boatman

Threats

The total number and area of salt marshes has been declining for many years. The main cause is enclosure, which removes the habitat from tidal inundation. This normally occurs in areas where the level is high enough and the area big enough for embankment to take place. This is usually achieved by the erection of a tidal barrier, sea wall or other structure designed to keep the sea out. This destroys the salt marsh as other uses, notably conversion to intensive agriculture, takes place.

In south east England a special type of enclosed saltmarsh develops its own value as a semi-natural habitat. Once enclosed the saltmarsh is left 'unimproved' with creeks, and other features retained with little interference. Grazing or cropping for hay are often the only uses and with traditional management it develops into a wildlife habitat of some significance. The term Coastal grazing marsh is used to describe this habitat.

Infilling for ports and harbours and other infrastructure completely destroys the habitat and with it any opportunities for restoration. Some areas outside the reclaimed zones can produce new marshes where relative sea level is falling or there is an abundant supply of new sediment and conditions suitable for growth of new plants. With climate change and associated sea level rise this is becoming much less prevalent and the process of 'coastal squeeze' occurs in many areas, especially around the margins of the southern North Sea.

Other threats include:

- Over-grazing;
- Embankment;
- Excessive agricultural use;
- Urbanization;
- Recreation;
- Coastal erosion;
- Pollution and industrial waste water;
- Invasion of *Spartina anglica* and *Erytrigia* (in eutrophic salt marshes).

Mangroves

Mangroves are the dominant ecosystems that line the coasts of subtropical and tropical coastlines around the world. Mangroves are survivors. With their roots submerged in water, mangrove trees thrive in hot, muddy, salty conditions that would quickly kill most plants. A series of adaptations allow these plants to not only live in these conditions but thrive, including a filtration system that keeps out much of the salt and a complex root system that holds the mangrove upright in the shifting sediments where land and water meet. Mangrove ecosystem also supports an incredible diversity of creatures, including some species unique to mangrove forests. And, as scientists are discovering, mangrove swamps are extremely important to our own well-being and to the health of the planet. There are four species of mangroves that range in size, characteristics, and preferred habitat. Mangroves act as natural buffers between the land and sea.

Mangroves absorb the force of strong waves that have been built up on the open seas. This is especially important during times of tropical storms. Additionally mangroves counteract erosion which waves cause over time. From a filtration standpoint mangroves neutralize sediment runoff from both natural and human activities, mangroves are often referred to as carbon sinks. Mangroves hold vital habitat to animals like birds that nest in their tight knit canopy. More importantly aquatic organisms that we as humans rely on are found in the tangled roots of the mangroves. Many juvenile fish from the nearby coral reefs, like snapper and grouper, come to take refuge in the tangled roots of the mangroves. Here the young fishes can grow with much less competition and predation. Mangrove lagoons are an important habitat for juveniles of many fish species, and can provide nursery areas for estuarine as well as reef fishes. Crustaceans such as shrimp, crabs, and lobster also take hold and grow in these ecosystems. Some smaller species of fish never leave the tangled root systems of the mangroves.

The area surrounding the edges of mangroves is the favorite hunting grounds for fish like the tarpon, rays, and lemon sharks. Another bounty for humans is the massive gastropods known as conch. Conchs are large slow moving snails that are highly sought after in the tropics of the Caribbean for their delectable meat. The mangrove wood itself has been collected for ages for fuel by those living in close proximity to the coastline. In recent years however mangroves are in decline as they are cleared for development, converted to aquiculture ponds, or wood chipped to aid in the production of rayon. This destruction displaces organisms that rely upon the mangroves for a rearing ground, hunting ground, and home. Additionally it puts its new human residents in danger of harsh waves and consequently the erosion that comes along with them.

Mangrove Species

Mangroves—forests of tropical trees and shrubs rooted in saltwater sediments between the coast and the sea—are crucial nurseries for coral reef fish. The tallest and most distinctive mangrove is the red mangrove (Rizophora mangle). Under optimal conditions this tree can reach 80 feet in height in some parts of the world, but on average reaches a height of 20 feet. The distinctive characteristic of this species is its reddish prop roots which can extend out of the water up to 3 feet. These roots are theprimary home for the many species of organisms that thrive in the ecosystem. The red mangrove is the most tolerant of turbulent water. This mangrove is the first specie to meet the aggressive waves from the open ocean. The black mangrove (Avicennia germinans) is smaller and grows closer to the shoreline in areas slightly higher in elevation, more susceptible to tidal fluctuations. In the times of low tide pencil-like projections, or pneumatophores, extend into the atmosphere to provide oxygen for the underwater root systems. This specie can reach over 65 feet in some locations, but on average it will reach heights up to 50 feet. This tree appears similar to that of the red mangrove with opposite leaf arrangement. However the presence of pneumatophores and lack of prop roots sets them apart. The white mangrove (Laguncularia racemosa) is found at even higher elevations where soils are least saturated. The most distinguishing feature is the lack of aerial roots. Unlike the black and red mangroves the white mangrove conceals its roots underground, with the exception of prolonged flooded areas where the tree will develop a sort of peg root to obtain oxygen. The fourth specie of mangrove is associated with the upland transitional zone is restricted in its range by its extreme sensitivity to cold. The buttonwood mangrove (Conocarpus erectus) is a shrub-like tree and is least like the other mangroves but still plays a vital role in the mangrove ecosystem being home to terrestrial animals and avian nesters.

Terrestrial Inhabitants

There is a wide range of mammals that visit the mangroves to hunt and forage for food. Some small like raccoons, opossums or agoutis, but there are also large animals that rely on these ecosystems. The very rare Florida panther was last seen in the everglades mangrove system, and due to habitat destruction it is estimated that roughly 50 individuals remain in the wild. The endemic key deer is only among the mangrove systems in the Florida Keys. In many other parts of the world monkeys are very common in mangroves, and in fact have been captured on video diving in the shallows collecting food. The greater flamingo and other special birds also visits these ecosystems were they forage in the shallows for tiny crustaceans. Many waterfowl species in North America spend their wintering months in and around the mangroves searching for food before their counter migration back north. To one particular sort of bird, mangrove canopies are among one of the few options for nesting. Pelagic birds, or sea birds, are birds that spend the majority of their lives at sea. These birds are highly efficient at staying aloft and use little energy when flying. They come to land rarely and often only to nest. To these birds the thick tangled canopy of the mangroves is ideal to build for nesting and rearing young. The seeds from a parent tree drift across the sea and rest upon an exposed section of sand, taking root and virtually creating a bird paradise. Black mangrove grows in the intertidal zone throughout the Gulf of Mexico. It will establish in nature from seed that floats and can travel some distance on the tides. These specialized birds, such as the magnificent frigate bird, can nest on an island free from predation and have an open buffet on either side to feed their young.

Aquatic Inhabitants

Among the tangle of the mangrove prop roots (Rizophora mangle) lays an absolute flourish of life. Nearby are beds of seagrass which are considered part of the mangrove ecosystem. The organisms from the mangroves interact with those of the seagrass beds. In the Caribbean, juvenile reef fish occupy the submerged prop roots of Rhizophora mangle and make frequent foraging runs into adjacent seagrass beds. Mangroves serve either as nurseries for juveniles or as feeding areas for transient fish and crustaceans. For many fish they start their lives in the adjacent sea grass beds, they migrate into the mangroves as they grow slightly larger, and then when they are just about fully mature they migrate across the flats to the coral reefs. This is known as ontogenic migration. It has been found that certain coral reefs are correlated to certain mangrove systems. Therefore if a mangrove forest is removed, the fish populations of the correlating coral reef are negatively affected. Just as waterfowl seem to have a sixth sense when it comes to direction, reef fish also seem to know where their desired destination is as the set off on their final migration to the coral reef where they will live out the rest of their lives. Migrating fishes can move directly between mangroves and reef habitats, passing over deep water and seagrass beds. The reasons for undertaking such a stage-structured lifecycle are varied and include: (i) requirements for different food sources as the organism grows, (ii) changing risks of predation with size, such that sheltered habitats, where predator foraging efficiency is low, are chosen when the organism occupies its smaller, more vulnerable stages, and (iii) a need to reproduce in habitats which offer the greatest dispersal or survival of larvae.

Mangroves as a Nursery

There is a wide range of fish species that live in the nursery of the mangroves. There are some fish species that live in the mangroves fulltime, some come and go with the seasons, and some just live in the mangroves until they reach maturity. Species that live here full time are a variety of gobies, spotted trout, pipefish, and others. These full time residents never grow too large for their habitat and therefore never have reason to leave. During the rainy season, the increased flow of freshwater results in the appearance of freshwater species, such as the predatory Florida gar. This increased flush of freshwater has been known to drive some species further off shore in search of higher saline waters. Among the resident species and seasonal guests are the juvenile fish species which migrate to the coral reefs once they have reached maturity. Of these juvenile reef fish are various species of grouper, snapper, jacks, schoolmasters, and others. Among the grouper species that live in these habitats is the endangered Goliath grouper. These spectacular fish can reach weights of over 800 lbs. when mature. Goliath grouper breed in mass swarms in the open ocean and their fry grow in the seagrass beds and then they migrate to the nearby tangle of the mangroves. Goliath grouper remained in mangrove habitats for 5–6yrs (validated ages from dorsal spine sections), then emigrated from mangroves at about 1.0 m total length.

In this mangrove nursery there are many species of fishes that flourish in a predator reduced environment. The larger predatory fish are too large to pursue their prey in the mangroves. While within the mangroves themselves there are very few predators, in the more open waters are predator fish that await any small wanderers that may stray away from the protection of the mangroves. Fish like the tarpon, snook, bone fish, red fish, rays, and lemon shark work with the tides. They are fairly large fish and need deeper water to enter the shallows surrounding the mangroves. Once these predators get into the shallows they are on a race against the clock to hunt as much as they can before the tides retreat and they become stranded. These fish are opportunistic feeders taking whatever they can fit in their mouths; just about everything is fair game. These fish don't have any natural predators in these waters. However they are highly sought after by the saltwater angler.

Other Mangrove Inhabitants

The ecosystem of the mangroves is not just home to fish species. Also calling home to these parts is a wide variety of invertebrates. The extensive root systems, muddy bottoms, and open waters are all home to invertebrates that are well adapted to the temperature and salinity fluctuations as well as the tidal influences common to mangroves. These wide array of organisms feed on leaf litter, detritus, plankton, algae, and other small animals. Barnacles, mollusks, shrimp, crabs, lobsters, jellyfish, tunicates, etc. can all be found among the roots of the mangrove. There is also a very ancient animal that can be found living in the mangroves, the earliest fossils of the horseshoe crab dates back to roughly 450 million years ago.

Horseshoe crabs are scavengers and may be found among mangroves feeding on algae, invertebrates, and dead organisms. They are especially adapted to low oxygen waters, possessing up to 200 book gills used for respiration. On the edges of the mangroves and out in the seagrass beds one can find colorful anemones with their feeding tentacles sent up to catch food, sea urchins crawling about munching algae, and large gastropods known as conch scooting about the bottom. These large slow moving snails are known for their iconic shells, and have been collected for years for their meat. Also commonly found in these waters and collected for their meat is the Caribbean spiny lobster. Their habitat usage is similar to that of the juvenile reef fish. Depending on their maturity they sought different habitat type. Largely influencing their preference for habitat was predation. Predation on newly settled juveniles was greater in seagrass and coral crevices than in mangrove prop roots, whereas the survival of larger juveniles was higher in mangroves and coral patch reefs than in seagrass. These results suggest that mangrove habitats may function as a nursery for juvenile spiny lobsters but that the use of this habitat depends on shelter characteristics and the isolation of islands.

Along with invertebrates and fishes, sea turtles are also associated with mangrove vegetation, at least point in their life. The Atlantic ridley sea turtle is commonly observed in the mangroves that line the bays of south Florida. The loggerhead and green sea turtles have been known to utilize the mangroves as juvenile nurseries as well, receiving protection from predation and ample access to a food rich environment. The green sea turtle and the Hawksbill sea turtles have been observed feeding on the roots of mangroves as well as other submerged vegetation. Another reptile known to the mangroves are saltwater crocodiles. Mangroves are particularly well suited for juveniles. All around the world species of water snakes can be found living in mangroves, as well as some saltwater tolerant amphibians. The most common of these amphibians is the giant toad (bufo marinus). The giant toad, or more commonly known as the cane toad, is native to Central and South America but has been introduced in the Caribbean and some Pacific Islands. Originally introduced to control the cane beetle it is now considered a pest. There are also two quite different aquatic animals that also are very common in and around mangroves. Feeding on the small fish species dolphin are quite common to mangroves. Feeding on seagrass and submerged vegetation manatees are also often seen swimming around the edges of mangroves along the coast of Florida. Another species of manatee that can be seen in coastal marine habitats is found in West Africa.

Erosion Resistance

Besides for crucial for wildlife habitat both above and below the surface, mangroves also serve a purpose to the land beyond them. Mangroves are very efficient at counteracting erosion. In many areas around the world mangroves are removed for various reasons, and in these areas the effect of erosion can be seen almost immediately. Mangroves act as a natural wave block from the strong waves that have built up on the open ocean. Even a small band of mangrove is an efficient blocking waves. Mangroves are extremely important in times of increased wave strength, such as during tropical storms. The best example on finds is the super-cyclone which occurred on the 29th October 1999 with a wind speed of 310 km hr-1 along the Orissa coast (India) and played havoc largely in the areas devoid of mangroves. On the contrary, practically no damage occurred in regions with luxuriant mangrove growth (K Kathiresan). This example along should be evidence enough to prove that mangroves are essential in protecting coastal lands. Ironically the highly populated areas most in need of protection from tropical storms are the area most likely to remove mangroves. A prime example of this would be Cancun, Mexico. Cancun is situated at the very tip of the Yucatan Peninsula, which is a relatively large piece of land situated in the middle of the Caribbean Sea. Given its location, the peninsula is highly susceptible to hurricane activity and with the removal of their mangroves they are rendered helpless to any strong tropical storm.

The logic for the removal is for more pleasing aesthetics, yet less aesthetically pleasing are the sandbags set in place to combat the waves and high water. Without the mangroves to hold sediment in place, even the gentlest of waves will take some amount of sediment out into the ocean. A well-known and credible institute has witnessed this first hand. The Smithsonian Institute operates a field station on a small island, 0.4 hectare in size, 14 miles off the cost of Dangriga, Belize. The Carrie Boy Field Station is located on the Meso-American Barrier Reef, and is devoid of mangroves. Do to this absence of mangroves the already small island shrinks every year, at an average decreasing rate of 2 inches. Along with keeping sediment in place, mangroves also filter the water passing through the sediment. "Black mangrove is valuable in restoring brackish and salt water marshes due to its ability to filter and trap sediments. Mangrove forests, which include black mangrove, have a high capacity as a sink for excess nutrients and pollutants. It also mixes well with other native plants to reduce wave energy".

Mangroves are so efficient at filtering nutrients from water that in some places they are being used on a municipal level to treat sewage. A mangrove wetland was constructed in Futian, Shenzhen, China to do just this. Three belts of mangroves were planted, housing three different species. The belts through which the water was fed held dimensions of 33 m in length, 3m in width, and 0.5 m in depth. Expectedly the removal of organic matter and nutrients were positively correlated with plant growth. The results indicated that mangroves could be used in a constructed wetland for municipal sewage treatment, providing post-treatment to remove coliforms was also included. In the natural setting filtration is a key as well, especially in areas where development is prevalent. The filtration of runoff is particularly important where the surrounding region contains seagrass or coral reefs. These habitats are particularly vulnerable to deterioration if water quality declines. If coral reefs and seagrass habitats were to be lost, numerous highly valuable ecosystem goods and services would also be lost.

Economic Importance

Mangroves are economically important as well. The value of mangrove ecosystem services worldwide has been estimated as an annual global flow of US$ 1,648 billion. The problem is when harvest exceeds production. Of 2500 households in 21 rural communities (about 13.000 people) near the Caeté estuary, 83% derive subsistence income, and 68% cash income through use of mangrove resources. Near this estuary on the northern Brazilian coast, everything is used in the ecosystem from the sea, to the land, and even the wood itself. The mangrove crab (Ucides cordatus) is collected and sold by 42% of households, and constitutes a main income source for 38%. Including processing and trading occupations, over half of the investigated populations depend on the mangrove crab for financial income. In addition to the mangrove crab in these regions, many households depend on the use and sale of the mangrove wood and bark. Not only in Brazil but virtually worldwide mangrove wood is collected on both the small scale and commercial scale. Mangrove wood is high in tannin therefore is very sought after for its durability in the timber industry. As with any wood harvest it is done both sustainably and not sustainably. In fact, mangroves can be very successfully managed for sustainable timber production – the Matang Mangroves in western Peninsular Malaysia are a classic example.

The management system, first started at the beginning of the 20th century, is based on a 30 year rotation. Small areas are clear-felled in a patchwork manner, and allowed to regrow before the next harvest. Mangrove wood burns with a high heat output and no smoke, making it ideal fuel wood. Coastal rural inhabitants have been using mangrove as fuel for many generations. Consequently these characteristics make it very attractive for mass collection coal production. One ton of mangrove wood is equivalent to 5 tons of Indian coal. Due to its durability the rayon industry also ravages the mangroves for their wood, as does the wood chip industry. These three industries are highly responsible for the destruction of mangroves. The pneumatophores can be used to make bottle stoppers and floats for fishing. A perhaps unexpected industry is the mangrove honey industry. During full bloom bees and other insects are found buzzing about mangroves as they are any patch of wild flowers. And in doing so collect pollen on their bodies which is later brought to the hive and put to use. Many people in foreign lands know of the sweet nectar produced and harvest this honey produced. For instance, the Sundarbans provide employment to 2000 people engaged in extracting 111 tons of honey annually and this accounts for about 90% of honey production among the mangroves of India. In Bangladesh, an estimated 185 tons of honey and 44.4 tons of wax are harvested each year in the western part of the mangrove forest.

Fisheries

Perhaps the most exploited, but economically profitable, parts of the mangrove ecosystem are the fisheries. On a subsistence level, fishing can be done sustainably and is quite crucial for some populations of people. On a local level many individuals dive for the Conch and keep themselves as well as sell them to local restaurants. But on a commercial level, animals are taken out of the mangroves for money and often at a rate at which the ecosystem cannot supply for the demand. Most common are shrimp and the various species of fish, although other crustaceans and shellfish are also sought after. Implementing the proper techniques the fisheries can be harvested from on a sustainable level. Most of the time the fishermen are often overwhelmed by greed, which ultimately leads to the down fall for the fishery. And when the ecosystem does not provide enough for what the demand is, aquaculture is established. Typically this means that the natural mangrove ecosystem is destroyed, and what remains is manipulated to better suit for a certain organism in order to gain cash profit. Consequently this only suits those particular organisms banishing the others from existing there. The food and other chemicals put into the water also adversely affect the rest of the surrounding waters around these man made ponds. The driving force is the high demand for shrimps in the developed countries and the governments of developing countries grossly undervaluing their mangroves.

Carbon Sequestration

Beyond the economical possibilities and the crucial wildlife habitat for so many varieties of organisms, is yet an even more important value of the mangrove. This is one that is important not only in small communities or even whole countries, but on a global scale. This unique characteristic is the mangroves ability to remove carbon from the atmosphere. Much like the wetlands of the inlands and the boreal forests of the temperate climates, mangrove systems are referred to as carbon sinks. The ocean itself does a great job through its direct air-to-sea exchange activity. This more or less is the action of waves capturing air and bringing the carbon into the ocean. With the combined oceanic air-to-sea exchange and the mangroves ability to seize atmospheric carbon, the two seize a huge amount of carbon. Measurements show that mangroves have the ability to seize around one and half tons of carbon per hectare per year. This is approximately equivalent to the amount of carbon a motor vehicle releases to the atmosphere each year (assuming each car uses approximately 2,500 liters of petrol per year). Indonesia has 4.5 million hectares of mangrove, and from this previous information we can conclude that the emission from 5 million cars can be erased by the mangrove ecosystem in Indonesian alone.

A high amount of carbon is stored in the upper layer of mangrove sediment, around 10 %. Each hectare of mangrove sediment consequently contains about 700 tons of carbon per every meter of depth, and digging up the sediment to create ponds for aquaculture results in the potential oxidation of 1/400 tons per hectare per year. Naturally only about half of this would become oxidized over the span of about ten years. The sequestration rate is then catalyzed to a rate of around 50 times the natural rate, thus eliminating the mangroves ability to absorb atmospheric carbon.

Intertidal Zone

The intertidal zone or "littoral zone" is the term used to describe the seashore which is covered during high tide and exposed during low tide, revealing a unique biome which survives under such fluctuating conditions. The size and location of the intertidal zone varies with the region and provides a habitat for a wide variety of plant and animal species. Moreover, the intertidal zone is characterized by unique environmental conditions, including the temperature, ecological factors, and microclimates.

Intertidal Zone Temperature

The temperature of the intertidal zone is extremely variable depending on the time of day and the status of the tide. It is common for the temperature to change by as much as 20°C within a time span as short as six hours. Such fluctuations in temperature impose a substantial physiological stressor on the various species which reside within the intertidal zone. For example, the Crassostrea gigas is a species of Pacific oyster which inhabits the intertidal zone. This species of oyster must adapt to massive temperature fluctuations. While various stress responses within the organism enable it to survive in this habitat, the metabolic rate is heavily influenced by the temperature, as is the ability of the organism to reproduce. Since the massive temperature fluctuation is poses such a stress, there is increasing concern that climate change will increase the temperatures of this region to a point which limits survival of the organisms which inhabit the intertidal zone.

Intertidal Zone Location

The location of the intertidal zone is any interface between the land and sea which is exposed during low tide and submerged at high tide. The intertidal zone can encompass sandy beaches, rocky shores, bays, and estuaries. The intertidal zone is divided into four distinct regions:

1. Lower Littoral Zone: The lower littoral zone is the area of the intertidal zone closest to the sea, and is submerged the majority of the time. This zone exhibits the greatest species diversity as there are less drastic fluctuations in salinity, temperature, and water coverage. In addition, the coverage of shallow water and waves are protective against harmful solar radiation and the severe temperatures that affect the shorelines during the day. As such, the species inhabiting this region are typically larger in size, greater in number, and more diverse than the other areas of the intertidal zone.

2. Mid-Littoral Zone: The mid-littoral zone is submerged half of the time, depending on whether it is high or low tide. Many species of plants and animals still inhabit this region; however, not to the same number or diversity as exhibited by the lower littoral zone.

3. Upper Mid-Littoral Zone: The upper mid-littoral zone is submerged only during high tide, and few plant and animal species are able to survive in this region. Since this region is exposed most of the time, most of the animals residing within this zone are mobile (e.g. crabs) or attached to the substrate (e.g. barnacles attached to rocks).

4. Splash Zone: The splash zone is the zone above the upper mid-littoral zone, which may experience splash from waves during high tide, but is never fully submerged during high tide.

Intertidal Zone Animals

Several animal species reside within the intertidal zone, each divided by the zones created by the tide. In the lower littoral zone, which remains submerged the majority of the time the organisms which inhabit this zone are generally larger, and protected from predation from the crashing waves. The organisms which reside in this zone include limpets, mussels, shrimp, crabs, tube worms, starfish, snails, and mollusks. In the mid-littoral zone, the organisms are equally exposed to water and land, depending on the tide. The animals residing within this zone include crabs, mussels, sea stars, sea sponges, snails, and limpets. The upper mid-littoral zone is only submerged under high tide, and thus, most organisms residing here are terrestrial. Organisms inhabiting this zone include, crabs, snails, mussels, and limpets. Finally, in the upper littoral zone, or spray zone, is dry the majority of the time, unless splashed with waves during high tide. Few organisms reside here, and include limpets, isopods, and barnacles.

Intertidal Zone Plants

Similar to the animal species residing within the intertidal zones, the vegetation which grows within each area is dictated by the relative exposure to water during the cycle of the tides. In the lower littoral zone, the plants are primarily adapted to water, including several species of seaweed, sea lettuce, sea palms, and green algae. These plants are typically larger and grow taller than those located in the other zones. In the mid-littoral zone, some species of seaweed and green algae reside; these plants are typically smaller and there is decreased species variation within this region. In the upper mid-littoral zone, few plants reside, due to the stressful fluctuations in salinity, temperature, and submersion. The species which do reside within this region include algae and some seaweed which survives in tidal pools during low tides. In the splash zone, vegetation is rare.

Rocky Shore

A rocky shore is an intertidal area that consists of solid rocks. It is often a biologically rich environment and can include many different habitat types like steep rocky cliffs, platforms, rock pools and boulder fields. Because of the continuously action of the tides, it is characterized by erosional features. Together with the wind, sunlight and other physical factors it creates a complex environment. Organisms that live in this area experience daily fluctuations in their environment. For this reason, they must be able to tolerate extreme changes in temperature, salinity, moisture and wave action to survive.

Rocky shore of the Costa Vicentina.

Zonation

Each region on the coast has a specific group of organisms that form distinct horizontal bands or zones on the rocks. The appearance of dominant species in these zones is called vertical zonation. It is a nearly universal feature of the intertidal zone.

Supratidal Zone

When the tide retreats, the upper regions become exposed to air. The organisms that live in this region are facing problems like gas exchange, desiccation, temperature changes and feeding. This upper region is called the supratidal or splash zone. It is only covered during storms and extremely high tides and is moistened by the spray of the breaking waves. Organisms are exposed to the drying heat of the sun in the summer and to extreme low temperatures in the winter. Because of these severe conditions, only a few resistant organisms live here. Common organisms are lichens. They are composed of fungi and microscopic algae living together and sharing food and energy to grow. The fungi trap moisture for both themselves and their algal symbiont. The algae on the other hand produce nutrients by photosynthesis. Green algae and cyanobacteria can also be found on the rocks of the North Atlantic coasts. They are capable of surviving on the moisture of the sea spray from waves. During the winter, they are found lower on the intertidal rocks. The algae growing higher on the rocks gradually die when the air temperature changes. At the lower edge of the splash zone, rough snails (periwinkles) graze on various types of algae. These snails are well adapted to life out of the water by trapping water in their mantle cavity or hiding in cracks of rocks. Other common animals are isopods, barnacles, limpets.

Intertidal Zone

Intertidal zonation: at low tide, the 3 typical intertidal zones can be seen.

The intertidal zone or littoral zone is the shoreward fringe of the sea bed between the highest and lowest limit of the tides. The upper limit is often controlled by physiological limits on species tolerance of temperature and drying. The lower limit is often determined by the presence of predators or competing species. Because the intertidal zone is a transition zone between the land and the sea, it causes heat stress, desiccation, oxygen depletion and reduced opportunities for feeding. At low tide, marine organisms face both heat stress and desiccation stress. The degree of this water loss and heating is determined by the body size and body shape. When body size increases, the surface area decreases so the water loss is reduced. Shape has a similar effect. Long and thin organisms dry up much faster than spherical organisms. Intertidal organisms can avoid overheating by evaporative cooling combined with circulation of body fluids. Higher-intertidal organisms are better adapted to desiccation than lower-intertidal organisms, because they encounter more hours of sun. The organisms are exposed directly to the air or they are enclosed in burrows. This results in oxygen depletion, so they can't get rid of their metabolic waste. A solution for this problem is to reduce the metabolic rate.

The intertidal zone can be divided in three zones:

- High tide zone or high intertidal zone: This region is only flooded during high tides. Organisms that you can find here are anemones, barnacles, chitons, crabs, isopods, mussels, sea stars and snails.
- Middle tide zone or mid-littoral zone: This is a turbulent zone that is (un)covered twice a day. The zone extends from the upper limit of the barnacles to the lower limit of large brown algae (e.g. *Laminariales, Fucoidales*). Common organisms are snails, sponges, sea stars, barnacles, mussels, sea palms and crabs.
- Low intertidal zone or lower littoral zone: This region is usually covered with water. It is only uncovered when the tide is extremely low. In contrast to the other zones, the organisms are not well adapted to long periods of dryness or to extreme temperatures. The common organisms in this region are brown seaweed, crabs, hydroids, mussels, sea cucumber, sea lettuce, sea urchins, shrimps, snails, tube worms.

Tidal pool in Santa Cruz.

Tidal pools are rocky pools in the intertidal zone that are filled with seawater. They are formed by abrasion and weathering of less resistant rock and scouring of fractures and joints in the shore platform. This leaves holes or depressions in where seawater can be collected at high tide. They can be small and shallow or deep. The smallest ones are usually found at the high intertidal zone, whereas the bigger ones are found in the lower intertidal zone. When the tide retreats, the pool becomes isolated. Because of the regular tides, the pool is not stagnant and new water regularly enters the pool. This is necessary to avoid temperature stress, salinity stress, nutrient stress. Pools that are located higher on the beach are not regularly renewed by tides. These pools are basically freshwater or brackish water communities. It has different characteristics in comparison with other coastal habitats. Several taxa are more abundant in pools than the surrounding environment. These taxa are members of the algae and gastropods. There is also a difference between high and low located pools for the composition. In low located pools, whelks, mussels, sea urchins and *Littorina littorea* are common. Periwinkles and *Littorina rudis* are found in high located pools. Other organisms that are commonly found in pools are flatworms, rotifers, cladocerans, copepods, ostracods, barnacles, amphipods, isopods, chironomid larvae and oligochaetes. Vertical zonation also has been documented in tidal pools.

Subtidal Zone

The subtidal zone or sublittoral zone is the region below the intertidal zone and is continuously covered by water. This zone is much more stable than the intertidal zone. Temperature, water pressure and sunlight radiation remain nearly constant. Organisms do not dry out as often as organisms higher on the beach. They grow much faster and are better in competition for the same niche. More essential nutrients are acquired from the water and they are buffered from extreme changes in temperature.

Problems and Adaptations

Air

Intertidal organisms are regularly exposed to air and water. Air differs physically from seawater in diverse and important features. This influences the ability to exchange gas and their overall thermal balance with the surrounding environment. Under water, organisms are generally buoyant, because of their lower density. In air, gravity induces retraction of tentacles and other feeding organs. It also makes the body less resistant. For this reason, organisms need supporting structures when they are exposed to air. Attachment and body changes are also required. When exposed to the air, organisms directly absorb solar radiation. The buffering capacity of water, because of the high rate of heat conductivity, disappears and the body temperature increases. In contrast to this, heat loss is much lower in air than in water. An adaptation to heating is the vaporization of internal water reserves.

Light

Sunlight is another parameter that influences the organisms. When there is too much sunlight, organisms dry out and the capacity to capture light energy can be weakened. The light that is not used or dissipated can cause damage to subcellular structures. Too little sunlight reduces the growth and reproduction of the organism, because photosynthesis is reduced. Algae can avoid absorbing too much light by changing the complement or amount of pigments they produce. They also can rearrange the pigmented organelles within their cells. When free radicals are produced from an excess of light, they can be scavenged and deactivated.

Temperature

The intertidal zone can experience extreme temperature changes. The organisms in this zone must be resistant to these changes to survive. Most of the marine organisms are ectothermic and need the warmth from the environment to survive. When the organisms are submerged, they are buffered against temperature changes, because the water is isothermal. When the organisms become exposed to the air, they can experience cool or warm temperatures. When the temperature is too low, the organisms must cope with physiological threats associated with cold stress. This can be the case in polar and temperate latitude coastal zones. The body fluids can then reach their freezing point and ice crystals develop. This causes damage to cell membranes and increasing the osmotic concentration of the remaining fluids. To avoid this cold stress, organisms can migrate to habitats that are more suitable. This can be a problem for sessile organisms. They can develop physiological and behavioral adaptations such as gaping shells (mussels). Some organisms have developed antifreeze proteins. Increasing the concentration of small osmolytes such as glycerol in the body fluids can decrease the freezing point. Another strategy is to control ice crystal formation. Organisms can control the speed and the exact location of the ice crystals.

When the ice formation is intracellular, it is lethal but extracellular ice formation can be tolerated. When the temperature is too high, heat stress appears. Heat stress accelerates rates of metabolic processes. This can be avoided by evaporative cooling combined with circulation of body fluids.

Salinity Stress

Salinity stress can occur in the external medium and in surface films. The concentration of the fluids determines whether or not the organism will lose water. When the osmolality of the cell is lower than the surrounding medium, the cell loses water from the internal fluids to the environment (hyperosmotic stress). When the intracellular osmolality is higher than the environment, there is an influx of water into the cell from the environment (hypoosmotic stress). Multicellular organisms respond to this salinity stress by compartmentalization. This buffers the cells from sharp changes in the osmotic environment. When the tissue has an immediate contact with the external medium, a solution can be to regulate intercellular osmotic pressure by actively excreting salts or water. Another solution is to change the internal osmolality. This can be done by incorporating ions or compatible solutes in the internal fluids.

Desiccation Stress

Organisms are threatened by desiccation during emersion at low tides or when they are positioned in the high intertidal zones. Deshydratation due to evaporative water loss is the most common mechanism. Highly mobile organisms can avoid the desiccation by migrating to a region that is more suitable. Less mobile organisms restrict various activities (reduced metabolism) and attach more firmly to the substrate. Physiological features to tolerate water loss are desiccation-resistant egg cases, reduction in water permeability of membranes, accumulation of metabolic end products, reduction of metabolic and developmental rates, maintenance of intracellular osmolytes and gene expression for production of protective macromolecules.

Predation

A wide variety of strategies to escape from predation exists. The first strategy is calcification. It makes it more difficult for the predator to eat these organisms. This strategy is applied by algae. It makes them tougher and less nutritious. A second one is the production of chemicals, usually produced as secondary metabolites. These chemicals can be produced all the time such as toxins, but other chemicals are only produced in response to stimuli (inducible defence). Another way to avoid predation is to have two distinct anatomical forms within one life cycle. This can be e.g. an alternation between a crusty form when the predator is present and a more delicate form (e.g. blade) when the predator is absent. Also the shape of the body can be a distinct evolutionary advantage. Bioluminescence is another strategy to avoid predators. Many intertidal and subtidal predators visually forage. The light is used for warning, blinding, making scare, misleading or attracting the predator. A commonly used form of protection against predation is camouflage. This can be visually or chemically. Visual camouflage means that the prey becomes invisible to the predator by using the same colors as the environment. Chemical camouflage is the passive adsorption of chemicals. The predator does not smell the prey anymore, because the smell is masked.

Wave Action

One way to protect organisms from waves is permanent attachment. But this strategy cannot be used by organisms that have to move to feed themselves. These organisms have to make a compromise between mobility and attachment. Another way to be protected is to burrow themselves into the sediment. But an alternative is to seek protected habitats. Attachment can be done by different structures. Bivalves usually use threads (byssal threads) to attach to rocky surfaces or to other organisms. But it can also be done by a foot. Another one is cementation. This is the case for bivalves such as oysters, scallops and some other forms. They lay on their side, with the lower valve cemented firmly to the bottom. This can be combined by reduction or enlargement of certain muscles.

Rocky Shore Morphology

Rocky shores constitute about 3⁄4 of the world coastline. They exist in very different forms. This diversity depends, among other things, on cliff mineralogy, lithology, tectonic history, climate, waves and tides. Some rocky coasts consist of steep plunging cliffs; this is observed when subaerial erosion is minor compared to marine erosion or in cases of volcanic activity or tectonic subsidence. Cliffs which are degraded by subaerial weathering processes are more gently sloping. Such more rounded (often vegetated) cliffs, are called "bluffs". A common coastal cliff morphology is the slope-over-wall profile: a rounded weathered top and a vertical wall below.

In many cases cliffs are fronted by shore platforms, which can be tens to many hundreds (or even thousands) of meters wide. These shore platforms result from cliff retreat; the outer edge marks the cliff position in the past – often millennia ago, in the case of erosion-resistant rock.

Wave attenuation on the shore platform reduces cliff recession rates. Cliff and shore platform both retreat by erosional processes; the rate of cliff retreat strongly depends on the width of the shore platform. An equilibrium profile is reached when platform retreat and cliff retreat occur at the same pace.

Cliff Erosion

Cliff Erosion

Cliff retreat depends on many factors: rock structure (massive, bedding, faulted, granular), rock lithology (permeability, solubility, hardness, jointing), hydraulic action (exposure to waves more or less loaded with debris, tides, currents) and subaerial weathering (by surface and subsurface runoff, freeze-thaw, chemical processes, biota). Hard rock cliffs retreat typically less than 1 cm/year, whereas soft cliffs (consisting of boulder clay, for example) retreat typically more than 1 m/year.

Cliff retreat is often episodic, through rock-fall and toppling, after undercutting of the cliff base by wave quarrying, abrasion and weathering. Wave quarrying is the process of rock dissection by breaking waves, which produce shock pressures and air compression in joints and fissures. Wave abrasion is the process of rock scraping and shattering by abrasive material (rock fragments) moved forth and back by wave-orbital motion. Rock weathering is due to the mechanical action of alternate drying and wetting and salt crystallization in fissures, but also due to chemical processes (dilution) and biotic activity.

Soft-rock cliffs may collapse by slumping or sliding events, which are often preceded by soil creep and triggered by rainwater infiltration. Debris fans at the cliff base after a slumping event are subsequently reworked by wave action; they are washed away or become part of the shore platform. Marine dissolution processes are an important cause of degradation of limestone cliffs, when acidic seawater (containing dissolved CO_2) and salt spray enter cracks, joints and fissures. Bio-erosion (by algae and by boring and grazing organisms) also contributes to erosion of calcareous rock. The resulting forms are termed "karst", featuring caves, arches and stacks.

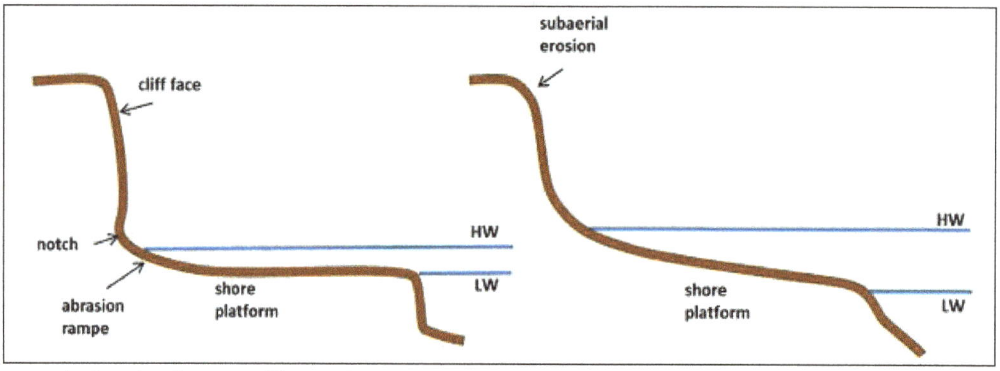

Typical rocky shore profiles. Left: Limestone cliff with low-tide terrace. Right: Hard rock cliff with sloping shore platform. Horizontal shore platforms are more characteristic for microtidal cliff coasts and sloping platforms for macrotidal cliff coasts.

Shore Platforms

Shore platforms can be either approximately horizontal, with a steep seaward edge, or gently seaward sloping. Tides modulate the frequency and intensity of wave attack at different levels of the shore platform. Sloping platforms prevail under macrotidal conditions and strong wave attack, whereas horizontal platforms are more common on micro-tidal coasts. High-tide terraces are typical for hard-rock coasts and low waves, low-tide terraces for limestone rocks. In the latter case, erosion resistance of the platform surface can be enhanced by carbonate precipitation and other cementing agents (silica, iron). Soft cliffs often have a beach at the base, with an intertidal veneer of sand and gravel.

Several processes contribute to erosion of shore platforms. The rate of platform down-wearing ranges from less than 1 mm/year to a few cm/year. Attack by waves armoured with rock fragments is an effective erosion mechanism, which polishes the platform surface. Fine abrasive material is washed away by waves and currents. However, platform surfaces usually present also many irregularities, with gullies, ponds and potholes scoured in faults, joints and in excavations of less resistant rock around hard outcrops. Chemical and physical weathering plays an important role, through the effects of alternate wetting and drying, salt crystallization and frost. Marine organisms accelerate platform erosion by grazing activity (surface scraping) and by boring of holes (by bivalves). Algae (e.g. kelp) can provide protection against erosion, but plant roots can also widen crevices and dislodge rock fragments.

Rocky shore platform, Holyhead, Wales (UK).

Sandy Shore

Sandy shores or beaches are loose deposits of sand, including some gravel or shells, that cover the shoreline in many places. They make up a large portion of the world's ice-free coastlines. Beaches serve as buffer zones or shock absorbers that protect the coastline, sea cliffs or dunes from direct wave attack. It is an extremely dynamic environment where sand, water and air are always in motion. Beaches also provide important coastal recreational areas for a many people. Fine-grained sand beaches tend to be gently sloping and quite flat.

Sandy beach in Middelkerke - Belgium.

Formation

Sandy beaches are soft shores formed by deposition of particles that have been carried by currents and waves. The transported material is in part derived from shore erosion, but the major part is generally derived from land and transported by rivers to the sea in the Holocene era or earlier. A more comprehensive introduction to beach formation is given in Coastal Hydrodynamics And Transport Processes. The two main types of beach material are quartz (silica) sands of terrestrial origin and carbonate sands of marine origin. The carbonate sand is weathered from mollusk shells and skeletons of other animals. Other material includes heavy minerals, basalt (volcanic origin) and feldspar.

Characteristics

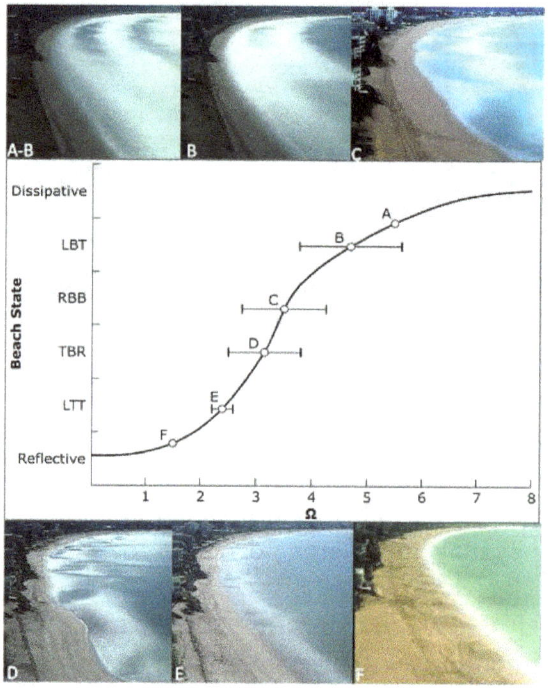

Different beach types based on morphodynamic scale.

The grain size of sand varies from very fine to very coarse. Quartz sands have a slightly lower density ($\sim 2.6 g.cm^{-3}$) than carbonate sands ($2.7-2.95 g.cm^{-3}$). The quartz particles are generally more rounded. Calcium carbonate particles sink more slowly in water due to their more irregular shapes, even if their density is higher.

Generic Name	Particle Diameter (mm)
Very coarse	1.0 to 2.0
Coarse	0.50 to 2.0
Medium	0.25 to 0.50
Fine	0.125 to 0.25
Very Fine	0.0625 to 0.125

Porosity is the volume of void space in the sand. It is the volume of water needed to saturate a given weight of dry sand. Most sands have a porosity of about 30 to 40 % of the total volume. The finer a sand the greater its porosity. Permeability is the rate of flow or drainage of water through the sand. Fine sands have lower permeabilities due to their smaller pore sizes. Penetrability is related to particle size and porosity. It can be important to the macrofauna. All species must be able to burrow into the substratum. To determine the penetrability, the proportion of clay and silt and the water content are very important.

The two basic beach types are dissipative and reflective. Together with the intermediate types, there are six major microtidal beach types. Beach types can also be based on the degree of exposure. This ranges from very sheltered over sheltered and exposed to very exposed.

Functioning and Adaptations

The intertidal beach zone is covered part of the day by water and is part of the day exposed to air. Swash and tides supply nutrients and food. When the tide retreats, waste products, eggs and larvae are taken away. Organisms living on sandy shores have adapted to this dynamic environment.

Burrowing on high-energy sandy beaches must be rapid and effective. This is because the animals must not be swept away by uprushing and downrushing water. In contrast with rocky shores, desiccation is not an overriding concern, because the animals can retreat into the substratum or below the water table. Intertidal filter-feeders cannot feed while the tide has retreated. Many species of the meiofauna use vertical tidal migrations through the sand column. Other species move up and down the beach with the tides. There is a difference between directional stimuli (such as light, slope of the beach, water currents) and nondirectional stimuli (such as disturbance of the sand, changes in temperature, hydrostatic pressure). Directional stimuli act as orientational signs, while nondirectional stimuli act as releasing factors. The predominant feeding types are filter-feeding and scavenging. Animals on low-energy sandy beaches adapt their respiration differently from those on surf-swept beaches. Some adaptations are an increased ventilation rate, an increased ventilation efficiency, reduced metabolic rate or other ways of energy economy. Many sheltered-shore animals are facultative anaerobes as an adaptation to ebb tides. Other animals in oxygenated surf-swept beaches are essentially aerobic. The majority of the intertidal animals have a high tolerance to variability in their environment, even exceeding what is necessary for survival in their particular habitats.

Some species bury themselves to escape high temperatures; others cool by evaporation, by entering the sea or by absorbing water from the substratum. Another problem for intertidal animals is the time of reproduction. There is variation in the number of eggs, the anatomy of the reproductive organs, the morphology of egg shells, times of breeding, mating behavior and developmental stages. Some species adapt by reproducing frequently (iteroparous) or by reproducing just once in a year (semelparous). There are also species that follow the lunar cycle to reproduce at the right time. To avoid predation, several behaviors have developed. The first one is deep burrowing. Another one is migration with the tide to escape predation. Crabs impress predators by holding their chelae open and aloft. According to circumstances, animals can modify their behavior. This is called phenotypic plasticity.

Several groups of vertebrates make use of sandy beaches for foraging, nesting and breeding. Turtles nest on the backshore of sandy beaches. Birds use the beach for foraging, nesting and roosting. Seals use several areas of the beach for nesting, molting, breeding and raising pups. Other terrestrial animals such as otters, baboons, raccoons, lions may descend onto the beach to forage.

Biota

The distribution and abundance of the sediment in fauna is mostly controlled by complex interactions between the physicochemical and biological properties of the sediment.

The physicochemical properties are:

- Grain size,
- Water content,
- Flushing rate of water through the sediment,
- Oxidation-reduction state,
- Dissolved oxygen,
- Temperature,
- Light,
- Organic content.

The biological properties are:

- Food availability and feeding activity.
- Reproductive effects on dispersal and settlement.
- Behavior that induces movement and aggregation.
- Intraspecific competition.
- Interspecific competition and competitive exclusion.
- Predation effects.

Most invertebrate phyla are represented on sandy beaches, either as interstitial forms or as members of the macrofauna. The macrofaunal forms are by far the better known. Some of them are typical of intertidal beaches and the surf zone, while others are more characteristic of sheltered sandbanks, sandy muds or estuaries and are less common on open beaches of pure sand.

Macrofauna

Macrofauna of sandy beaches are often abundant and, in some cases, attain exceptionally high densities. Their main feature is the high degree of mobility displayed by all species. These animals may vary from a few mm to 20 cm in length. The macrofauna community consists of organisms too large to move between the sand grains. The macrofauna of sandy beaches includes most major invertebrate taxa although it has been recognised that molluscs, crustaceans and polychaetes are the most important. There is a tendency for crustaceans to be more abundant on tropical sandy beaches or more exposed beaches and molluscs to be more abundant on less exposed and on temperate beaches although there are many exceptions of this and polychaetes are sometimes more abundant than either of these taxa. Generally crustaceans dominate the sands towards the upper tidal level and molluscs the lower down level. Physical factors, primary wave action and particle size of the sand largely determine distribution and diversity of the invertebrate macrofauna of sandy beaches. Food input and surf-zone productivity may determinate the population abundance.

Meiofauna

In contrast to the wave-swept beach surface inhabited by most of the macrofauna, the interstitial system is truly three-dimensional, often having great vertical extent in the sand. The porous system averages about 40% of the total sediment volume. Its inhabitants include small metazoans forming the meiofauna, protozoans, bacteria and diatoms. The meiofauna is defined as those metazoan animals passing undamaged though 0.5 to 1.0 mm sieves and trapped on 30 mm screens. On most beaches the interstitial fauna is rich and diverse, even exceeding in some cases the macrofauna in biomass. The dominant taxa of sandy beach meiofauna are nematodes and harpacticoid copepod with other important groups including turbellarians, oligochaetes, gastrotrichs, ostracods and tardigdades.

Insects

Terrestrial insects and vertebrates are frequently ignored in accounts of sandy beaches. These animals are usually a conspicuous component of the ecosystems, often rivalling the aquatic macrofauna in terms of biomass and having a significant impact on the system with regard to predation and scavenging.

Estuary

Estuary is a partly enclosed coastal body of water in which river water is mixed with seawater. In a general sense, the estuarine environment is defined by salinity boundaries rather than by geographic boundaries. Estuaries are places where rivers meet the sea and may be defined as areas where salt water is measurably diluted with fresh water. On average, estuaries are biologically more productive than either the adjacent river or the sea, because they have a special kind of water circulation that traps plant nutrients and stimulates primary production. Fresh water, being lighter than salt water, tends to form a distinct layer that floats at the surface of the estuary. At the boundary between fresh and salt water, there is a certain amount of mixing caused by the flow of fresh water over salt and by the ebb and flow of tides. Additional mixing may be caused from time to time by strong winds and by internal waves that are propagated along the interface between fresh and salt water.

Many coastal features that are designated by other names are in fact estuaries. For example, various coastal embayments, such as Chesapeake Bay and Galveston Bay in the United States, also are estuaries, because fresh and salt water undergo considerable mixing. Moreover, most of the world's submerged fjord systems (such as Scoresby Sund, Greenland) and large semi-enclosed tidal flat regions and coastal marshes (such as the Waddenzee area of The Netherlands) are estuaries. In addition, estuaries include river mouths (as in the case of the Mississippi River, Louisiana), structural basins (San Francisco Bay, California), and the bodies of water behind spits (Hurst Castle spit, England) and barrier beaches (Ninety Mile Beach, Australia).

Geology and Geomorphology

The geologic processes that form an estuary are extremely complex and varied, but it is clear that the existence of an estuary is largely dependent on the position of sea level relative to the freshwater discharge. If sea level were lowered, the estuarine zone would migrate seaward at the interface of the marine water and the edge of the newly exposed land area. Such migration has occurred as a consequence of Earth's several glaciations. For each glaciation, the primary source of moisture has been the oceans. Whenever sea level fell, the estuarine environment at the continental margin was forced to migrate seaward.

About 18,000 years ago the Wisconsin Glacial Stage attained its maximum, and glacial melting began. The seas rose, forcing the estuarine environment to migrate back up the continental shelf. During the period of lowered sea level, some rivers had become entrenched in the continental shelf and deepened their valleys, which were soon flooded by the rising marine waters, forming a typical drowned river estuary. In areas such as Norway and parts of the coast of British Columbia, Canada, valley glaciers had deepened river valleys. These narrow drowned glacial valleys became the modern fjord estuaries as sea level rose.

The geomorphology of an estuarine basin is usually developed by one of three agents: fluvial or glacial erosion, fluvial and marine deposition, or tectonic activity. The last of these involves the downfaulting of a coastal area or the broad local subsidence of a stretch of coastline, as in the case of San Francisco Bay.

Primary Productivity

The high level of plant production in estuaries supports a correspondingly high level of production of invertebrate animals and fish. Estuaries often contain beds of shellfish such as mussels and oysters and large populations of shrimps and crabs. Fish such as plaice and flounders are common. Other species use the estuaries as nursery grounds. Organisms in early stages of development enter the salt wedge at the seaward end and are carried up the estuary by the bottom currents. Juveniles find abundant food as well as protection from predators in the mangrove forests, salt marshes, or sea grass beds that line the estuary. Later they may migrate to the open ocean to continue their growth and development. Other species pass through the estuaries in the course of their migrations. For example, salmon migrate from the sea to the rivers to spawn, and the young fish later migrate back to the sea. Eels migrate in the opposite direction, breeding in the sea but returning to fresh water as juveniles.

Commercial Importance

Estuaries have long been important as harbour sites and centres of commerce. Some of the world's oldest continuous civilizations have flourished in estuarine environments, such as the lower reaches of the Tigris and Euphrates rivers, the Indus River, the Po River delta region of Italy, the Nile delta, the Ganges delta, and the lower Huang He (Yellow River) valley. Developing civilizations soon discovered that the logical site for commercial seaports was the seaward most point of the major river systems. Such cities as London (River Thames), New York City (Hudson River), Montreal (St. Lawrence River), Hamburg (Elbe River), and Bordeaux (Gironde estuary) developed on estuaries and became important centres of commerce. The proximity of estuaries to large urban centres, however, has made them particularly vulnerable to contamination by sewage and industrial effluents. The characteristic circulation that serves to trap natural plant nutrients may also retain high concentrations of pollutants.

In addition, many estuaries are important sites for aquaculture. There is a long history of mussel culture along the coast of Spain, and Norwegian fjords are much used for salmon culture. In Southeast Asia artificial ponds are created in mangrove forests and are used to culture shrimp.

Classification of Estuaries on the Basis of Water Balance

On the basis of the definitions, and in terms of their water balance, estuaries can be classified as three types: positive, inverse and low-inflow estuaries figure). Positive estuaries are those in which freshwater additions from river discharge, rain and ice melting exceed freshwater losses from evaporation or freezing and establish a longitudinal density gradient. In positive estuaries, the longitudinal density gradient drives a net volume outflow to the ocean, as denoted by stronger surface outflow than near-bottom inflow, in response to the supplementary fresh-water. The circulation induced by the volume of freshwater added to the basin is widely known as 'estuarine' or 'gravitational' circulation.

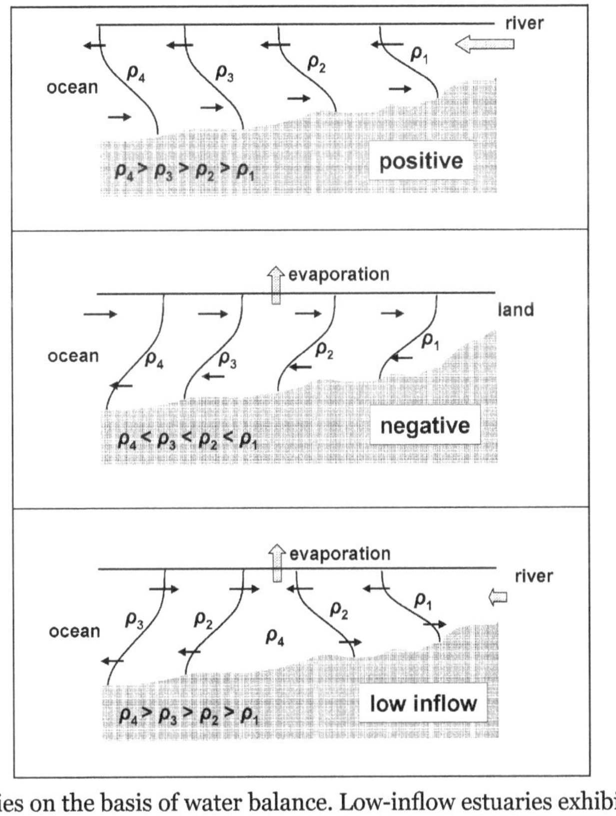

Types of estuaries on the basis of water balance. Low-inflow estuaries exhibit a "salt plug".

Inverse estuaries are typically found in arid regions where freshwater losses from evaporation exceed freshwater additions from precipitation. There is no or scant river discharge into these systems. They are called inverse, or negative, because the longitudinal density gradient has the opposite sign to that in positive estuaries, i.e., water density increases landward. Inverse estuaries exhibit net volume inflows associated with stronger surface inflows than near-bottom outflows.

Water losses related to inverse estuaries make their flushing more sluggish than positive estuaries. Because of their relatively sluggish flushing, negative estuaries are likely more prone to water quality problems than positive estuaries.

Low-inflow estuaries also occur in regions of high evaporation rates but with small (on the order of a few m^3/s) influence from river discharge. During the dry and hot season, evaporation processes may cause a salinity maximum zone (sometimes referred to as a salt plug) within these low-inflow estuaries. Seaward of this salinity maximum, the water density decreases as in an inverse estuary. Landward of this salinity maximum, the water density decreases, as in a positive estuary. Therefore, the zone of maximum salinity acts as a barrier that precludes the seaward flushing of riverine waters and the landward intrusion of ocean waters. Because of their weak flushing in the region landward of the salinity maximum, low-inflow estuaries are also prone to water quality problems.

Classification of Estuaries on the Basis of Geomorphology

Estuaries may be classified according to their geomorphology as coastal plain, fjord, bar-built and tectonic. Coastal plain estuaries, also called drowned river valleys, are those that were formed as a result of the Pleistocene increase in sea level, starting ~15,000 years ago. Originally rivers, these estuaries formed during flooding over several millennia by rising sea levels. Their shape resembles that of present-day rivers, although much wider. They are typically wide (on the order of several kilometres) and shallow (on the order of 10 m), with large width/depth aspect ratios. Examples of these systems are Chesapeake Bay and Delaware Bay on the eastern coast of the United States.

Fjords are associated with high latitudes where glacial activity is intense. They are characterized by an elongated, deep channel with a sill. The sill is related to a moraine of either a currently active glacier or an extinct glacier. In the sense of the glacier activity, it could be said that there are riverine and glacial fjords. Riverine fjords are related to extinct glaciers and their main source of buoyancy comes from river inputs. They are usually found equator ward of glacial fjords. Glacial fjords are found in high latitudes, pole ward of riverine fjords. They are related to active glaciers and their main source of buoyancy is derived from melting of the glacier and of snow and ice in mountains nearby. Fjords are deep (several hundreds of meters) and narrow (several hundreds of meters) and have low width/depth aspect ratios with steep side walls. Fjords are found in Greenland, Alaska, British Columbia, Norway, New Zealand, Antarctica and Chile.

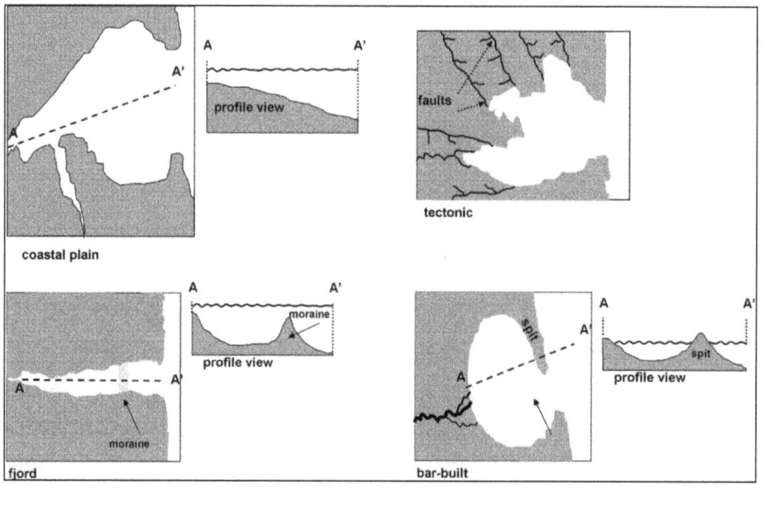

Bar-built estuaries, originally embayment's, became semi enclosed because of littoral drift causing the formation of as and bar or spit between the coast and the ocean. Some of these bars are joined to one of the headlands of a former embayment and display one small inlet (on the order of a few hundred meters) where the estuary communicates with the ocean. Some other sand bars may be detached from the coast and represent islands that result in two or more inlets that allow communication between the estuary and the ocean. In some additional cases, sand bars were formed by rising sea level. Examples of bar-built estuaries abound in subtropical regions of the Americas (e.g. North Carolina, Florida and northern Mexico) and southern Portugal.

Tectonic estuaries were formed by earthquakes or by fractures of the Earth's crust, and creases that generated faults in regions adjacent to the ocean. Faults cause part of the crust to sink, forming a hollow basin. An estuary is formed when the basin is filled by the ocean. Examples of this type of estuary are San Francisco Bay in the United States, Manukau Harbour in New Zealand, and Guaymas Bay in Mexico.

Classification of Estuaries on the Basis of Vertical Structure of Salinity

According to water column stratification or salinity vertical structure, estuaries can be classified as salt wedge, strongly stratified, weakly stratified or vertically mixed. This classification considers the competition between buoyancy forcing from river discharge and mixing from tidal forcing. Mixing from tidal forcing is proportional to the volume of oceanic water entering the estuary during every tidal cycle, which is also known as the tidal prism. Large river discharge and weak tidal forcing results in salt wedge estuaries such as the Mississippi (USA), Rio de la Plata (Argentina), Vellar (India),Ebro (Spain), Pánuco (Mexico), and Itajaí-Açu (Brazil). These systems are strongly stratified during flood tides, when the ocean water intrudes in a wedge shape. Some of these systems lose their salt wedge nature during dry periods. Typical tidally averaged salinity profiles exhibit a sharp pycnocline (or halocline), with mean flows dominated by outflow throughout most of the water column and weak inflow in a near-bottom layer. The mean flow pattern results from relatively weak mixing between the inflowing ocean water and the river water.

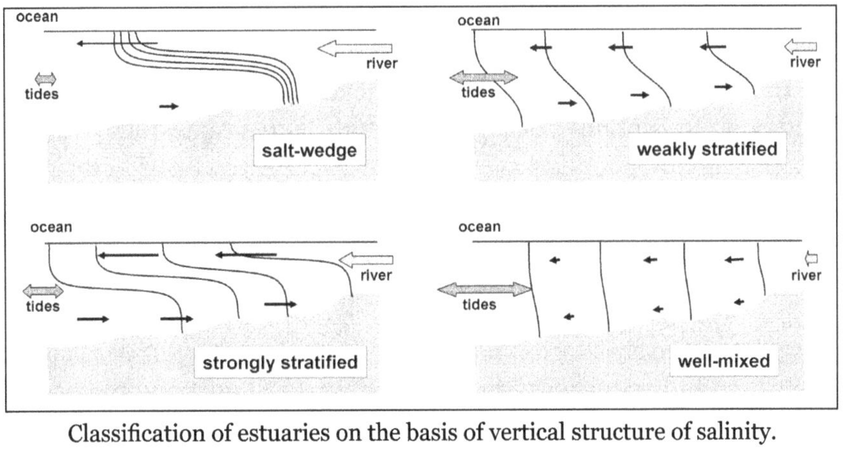

Classification of estuaries on the basis of vertical structure of salinity.

Moderate to large river discharge and weak to moderate tidal forcing resultin strongly stratified estuaries. These estuaries have similar stratification to salt wedge estuaries, but the stratification remains strong throughout the tidal cycle as in fjords and other deep (typically >20 m deep) estuaries. The tidally averaged salinity profiles have a well-developed pycnocline with weak vertical variations above and below the pycnocline. The mean tidal flow exhibits well-established outflows and inflows, but the inflows are weak because of weak mixing with freshwater and weak horizontal density gradients.

Weakly stratified or partially mixed estuaries result from moderate to strong tidal forcing and weak to moderate river discharge. Many temperate estuaries, such as Chesapeake Bay, Delware Bay and James River (all in the eastern United States) fit into this category. The mean salinity profile either has a weak pycnocline or continuous stratification from surface to bottom, except near the bottom mixed layer. The mean exchange flow is most vigorous (when compared to other types of estuaries) because of the mixing between riverine and oceanic waters.

Strong tidal forcing and weak river discharge result in vertically mixed estuaries. Mean salinity profiles in mixed estuaries are practically uniform and mean flows are unidirectional with depth. In wide (and shallow) estuaries, inflows may develop on one side across the estuary and outflow on the other side, especially during the dry season. Parts of the lower Chesapeake Bay may exhibit this behavior in early autumn. In narrow well-mixed estuaries, inflow of salinity may only occur during the flood tide because the mean flow will be seaward. Examples of this type of estuary are scarce because, under well-mixed conditions, the mean (as in the tidally averaged sense) flow will most likely be driven by wind or tidal forcing.

It is essential to keep in mind that many systems may change from one type to another in consecutive tidal cycles, or from month to month or from season to season, or from one location to another inside the same estuary. For instance, the Hudson River, in the eastern United States, changes from highly stratified during neap tides to weakly stratified during spring tides. The Columbia River, in the western United States, may be strongly stratified under weak discharge conditions and similar to a salt-wedge estuary during high discharge conditions.

Classification of Estuaries on the Basis of Hydrodynamics

A widely accepted classification of estuaries was proposed by Hansen and Rattray on the basis of estuarine hydrodynamics. This classification after acquiring a basic understanding of estuarine hydrodynamics. This classification is anchored in two hydrodynamic non-dimensional parameters: (a) the circulation parameter and (b) the stratification parameter. These parameters refer to tidally averaged and cross-sectionally averaged variables. The circulation parameter is the ratio of near-surface flow speed us to sectionally averaged flow U_f. The near-surface flow speed is typically related to the river discharge and, for the sake of argument, on the order of 0.1 m/s. The depth-averaged flow U_f is typically very small, tending to zero, in estuaries of vigorous water exchange because there will be as much net outflow as net inflow. In estuaries with weak net inflow, such as well-mixed and salt-wedge systems, the depth-averaged flow will be similar in magnitude to the surface outflow. Therefore, the circulation parameter is >10 in estuaries with vigorous gravitational circulation and close to 1 in estuaries with unidirectional net outflow. In general the greater the circulation parameter, the stronger the gravitational circulation.

The other non-dimensional parameter, the stratification parameter, is the ratio of the top-to-bottom salinity difference ∂S to the mean salinity over an estuarine cross- section S_0. A ratio of 1 indicates that the salinity stratification (or top-to-bottom difference) is as large as the sectional mean salinity. For instance, if an estuary shows a sectional mean salinity of 20, for it to exhibit a stratification parameter of 1 it must have a very large stratification (on the order of 20). In general, estuaries will most often have stratification parameters <1. The weaker the water column stratification, the smaller the stratification parameter will be.

The two parameters described above can be used to characterize the nature of salt transport in estuaries. The contribution by the diffusive portion (vs the advective portion) of the total salt flux into the estuary can be called v. The parameter v may oscillate between 0 and 1. When v is close to 0, up-estuary salt transport is dominated by advection, i.e., by the gravitational circulation. In this case, mixing processes are weak, as in a highly stratified estuary (fjord). When v approaches 1, the total salt transport is dominated by diffusive processes (e.g. tidal mixing), as in unidirectional net flows. The parameter v may be portrayed in terms of the stratification and circulation parameters. This diagram shows that salt transport is dominated by advective processes under high gravitational circulation or strongly stratified conditions. It also shows that diffusive processes dominate the salt flux at low circulation parameter (unidirectional net flows) regardless of the stratification parameter. Between those two extremes, the salt transport has contributions from both advective and diffusive processes. The more robust the stratification and circulation parameters, the stronger the contribution from advective processes to the total salt flux will be.

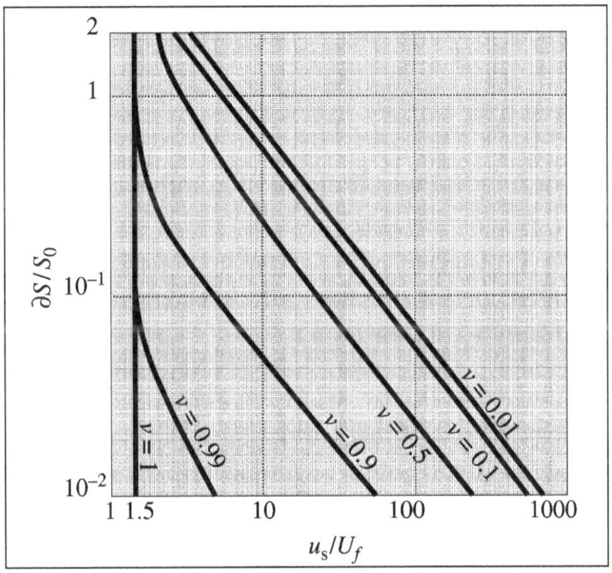

Diffusive saltflux fraction in the stratification/circulation parameter space.

These concepts can be used to place estuaries in the parameter space of the circulation and stratification parameters. The lower-left corner of the parameter space describes well-mixed estuaries with unidirectional net outflows, i.e., seaward flows with no vertical structure or type 1 estuaries. These are well- mixed estuaries, type 1a, implying strong tidal forcing and weak river discharge (or large tidal prisms relative to freshwater volumes). There are also estuaries with depth-independent seaward flow but with highly stratified conditions. These type 1b estuaries have large river discharge compared to tidal forcing. In type 1 estuaries, in general, the upstream transport of salt is overwhelmingly dominated by diffusive processes (v ≈ 1, figure).

Classification of estuaries according to hydrodynamics, in terms of the circulation and stratification parameters: (A) Type 1 estuaries show no vertical structure in net flows; in type 2 estuaries, the net flows reverse with depth; type 3 estuaries exhibit strong gravitational circulation; and type

4 estuaries are salt wedge. (B) Includes lines of diffusive salt flux showing the dominance of advective salt flux for type 3 estuaries and diffusive flux for type 1.

Type 2 estuaries are those where flow reverses at depth, and include most temperate estuaries. These systems feature well-developed gravitational circulation and exhibit contributions from advective and diffusive processes to the upstream salt transport (0.1 < ν < 0.9). Type 2a estuaries are well-mixed or weakly stratified and type 2b estuaries are strongly stratified. Note that strongly stratified and weakly stratified estuaries of type 2 may exhibit similar features in terms of the relative contribution from diffusive processes to the upstream salt transport.

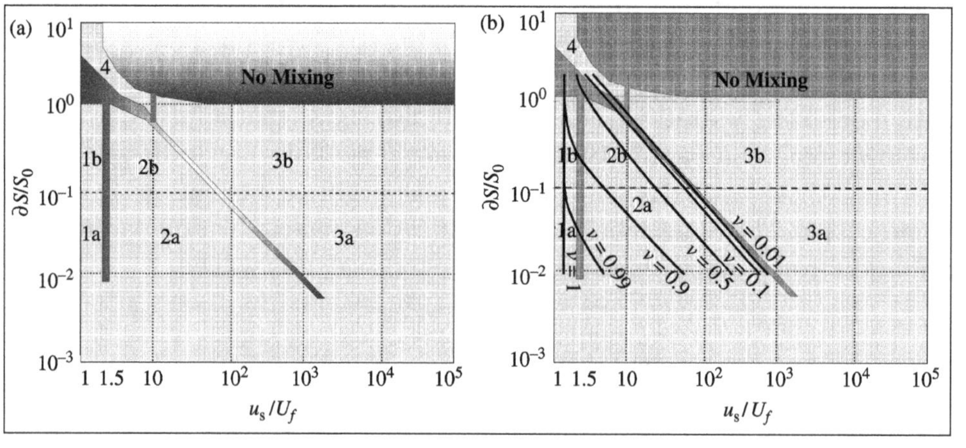

Classification of estuaries according to hydrodynamics, in terms of the circulation and stratification parameters. (A) Type 1 estuaries show no vertical structure in net flows; in type 2 estuaries, the net flows reverse with depth; type 3 estuaries exhibit strong gravitational circulation; and type 4 estuaries are salt wedge. (B) Includes lines of diffusive salt flux showing the dominance of advective salt flux for type 3 estuaries and diffusive flux for type 1.

Type 3 estuaries are associated with fjords, where gravitational circulation is well established: strong surface outflow and very small depth-averaged flows, typical of deep basins. This flow pattern results in large values (>100–1000) of the circulation parameter. Type 3a estuaries are moderately stratified and type 3b are highly stratified. The peculiarity about these systems is that the upstream transport of salt is carried out exclusively by advective processes (ν < 0.01, figure).

Finally, type 4 estuaries exhibit seaward flows with weak vertical structure and highly stratified conditions as in a salt-wedge estuary. In type 4 estuaries, the diffusive fraction ν lines tend to converge, which indicates that in type 4 estuaries, salt transport is produced by both advective and diffusive processes. In the Hansen– Rattray diagram, it is noteworthy that some systems will occupy different positions in the parameter space as stratification and circulation parameters change from spring to neap tides, from dry to wet seasons, and from year to year.

Analogous to the classification of estuaries in terms of the two non-dimensional parameters discussed above, estuarine systems can also be classified in terms of the lateral structure of their net exchange flows. The lateral structure may be strongly influenced by bathymetric variations and may exhibit vertically sheared net exchange flows, i.e., net outflows at the surface and near-bottom inflows, or laterally sheared exchange flows with outflows over shallow parts of a cross-section and inflows in the channel. The lateral structure of exchange flows may ultimately depend on competition between the Earth's rotation (Coriolis) and frictional effects, as characterized by the vertical Ekman number (Ek). But the lateral structure of exchange flows may also depend on the Kelvin number (Ke), which is the ratio of the width of the estuary to the internal radius of deformation.

The Ekman number is a non-dimensional dynamical depth of the system. Low values of Ek imply that frictional effects are restricted to a thin bottom boundary layer (weak frictional, nearly geostrophic conditions), while high values of Ek indicate that friction affects the entire water column. The lateral structure of density-driven exchange flows may be described in terms of whether the flows are vertically sheared or unidirectional in the deepest part of the cross-section. Under low Ek (< 0.001, i.e., < −3 in the abscissa of figure), the lateral structure of exchange flows depends on the dynamic width of the system (figure). In wide systems (Ke > 2, i.e., > 0.3 in the ordinate of figure), outflows and inflows are separated laterally according to the Earth's rotation, i.e., the exchange flow is laterally sheared. In narrow systems (Ke < 1, i.e., < 0 in the ordinate of figure) and low Ek (still < 0.001, i.e., < −3 in the abscissa of figure), exchange flows are vertically sheared. In contrast, under high Ek (> 0.3, i.e., > −0.5 in the abscissa of figure) and for all Ke, the density-driven exchange is laterally sheared independently of the width of the system. Finally, under intermediate Ek (0.01 < Ek < 0.1, i.e., between −2 and −1 in the abscissa of figure), the exchange flow is preferentially vertically sheared but exhibiting lateral variations.

The main message is that under weak friction (Ek < 0.01), both depth and width are important to determine whether the density-driven exchange is vertically or horizontally sheared. This is illustrated by the fact that the contour values in the entire region of Ek < 0.01 (i.e., < −2 in the abscissa) in figure vary with both Ek and Ke. In contrast, under Ek > 0.01 the depth is the main determinant as to whether the exchange is vertically or horizontally sheared. This is shown by the fact that the contour values in Figure vary mostly with Ek but very little with Ke. A future challenge of this approach is to determine the variability of a particular system in the Ek vs Ke parameter space. It is likely that an estuary will describe an ellipse of variability in this plane from spring to neaps and from wet to dry seasons, or from year to year.

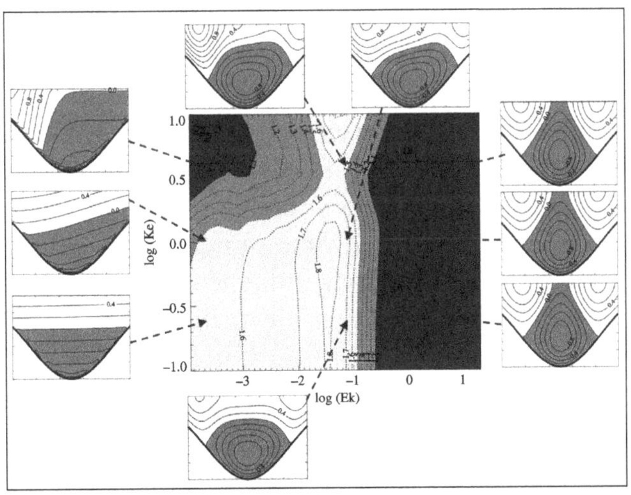

Classification of Estuarine Exchange on the Basis of *Ek* and *Ke*

The subpanels appearing around the central figure denote cross-sections, looking into the estuary, of exchange flows normalized by the maximum inflow. Inflow contours are negative and shaded in orange. The vertical axis is non- dimensional depth from 0 to 1 and the horizontal axis is non-dimensional width, also from 0 to 1. The central figure illustrates contours of the difference between maximum outflow and maximum in flow over the deepest part of the channel and for different values of Ek and Ke. Note that the abscissa and ordinate represent the logarithm of Ek and Ke. Blue-shaded contour regions denote net inflow throughout the channel, i.e., laterally sheared exchange flow as portrayed by the subpanels whose arrows point to the corresponding Ek and Ke in the blue regions. Yellow-shaded contour regions illustrate vertically sheared exchange in the channel as portrayed by the subpanels whose arrows point to the corresponding Ek and Ke in the yellow regions. Green regions represent vertically and horizontally sheared exchange flow, similar to the second subpanel on the left, for log(Ke) = 0 and log(Ek)~−3.7.

All of the classifications depend on diagnostic parameters that require substantial information about the estuary, i.e., on dependent variables. In addition, they do not take into account the effects of advective accelerations, related to lateral circulation that may be of the same order of magnitude as frictional effects.

Lagoon

Lagoons are highly productive coastal features that provide a range of natural services that society values. Their setting within the coastal landscape leaves them especially vulnerable to profound physical, ecological, and associated societal disturbance from global climate change. Expected shifts in physical and ecological characteristics range from changes in flushing regime, freshwater inputs, and water chemistry to complete inundation and loss and the concomitant loss of natural and human communities. Therefore, managing coastal lagoons in the context of global climate change is critical. Although management approaches will vary depending on local conditions and cultural norms, all management scenarios will need to be nimble and to make full use of the spectrum of values through which society views these unique ecosystems. We propose that this spectrum includes pragmatic, scholarly, aesthetic, and tacit categories of value. Pragmatic values such as fishery or tourism revenue are most easily quantified and are therefore more likely to be considered in management strategies. In contrast, tacit values such as a sense of place are more difficult to quantify and therefore more likely to be left out of explicit management justifications. However, tacit values are the most influential to stakeholder involvement because they both derive from and shape individual experiences and beliefs. Tacit values underpin all categories of social values that we describe and can be expected to have a strong influence over human behavior. The articulation and inclusion of the full spectrum of values, especially tacit values, will facilitate and support nimble adaptive management of coastal lagoon ecosystems in the context of global climate change.

Coastal lagoons support a range of natural services that are highly valued by society, including fisheries productivity, storm protection, tourism, and others. We review how global climate change (GCC) will affect the physical structure, ecological characteristics, and social values associated with lagoons, with special emphasis on the U.S. Atlantic and Gulf of Mexico coasts.

This follows similar ecosystem-specific assessments of the impacts of climate change in marine and coastal environments such as those of tropical coral reefs, Amazonian forests, and city ecosystems. We summarize the key characteristics of lagoons in terms of their physical and ecological properties. The social values of lagoons and offer a nomenclature to frame discussion about these values. Although there is considerable literature on the physical and ecological dimensions of lagoons, there is much less literature focused on their economic and social values. Finally, we discuss how the physical, ecological, and social dimensions of lagoons are likely be impacted by GCC in the coming decades. Examples are presented from four geographically and culturally distinct lagoon ecosystems to illustrate the range of possible impacts of GCC.

Climate change is only one of many sources of disturbance to lagoon ecosystems, and these disturbances occur concurrently at multiple temporal and spatial scales. Whereas some climate change stressors manifest slowly over decades and on regional and continental spatial scales, local, site-specific stressors can occur rapidly and cause significant impacts to lagoons. Land-use change, freshwater withdrawal from ground and surface water sources, sedimentation, point and nonpoint water pollution, shoreline hardening, and overfishing are examples of anthropogenic stressors that can have profound and sudden impacts on coastal ecosystems. Interactions among the host of local stressors and GCC can also be important, but interactions are complex and highly site specific. It is a major challenge for coastal researchers and managers to understand the interactions among these multiscale stressors. These important issues are beyond the scope of our expection. Instead, we restricted our focus to examine long-term social and ecological effects of GCC-induced changes to lagoon ecosystems.

Lagoon Ecosystems

Physical Characteristics of Lagoons

A coastal lagoon is a "shallow coastal water body separated from the ocean by a barrier, connected at least intermittently to the ocean by one or more restricted inlets". For our purposes, we restricted our survey to lagoons separated from the ocean by a depositional land barrier, excluding coral reef lagoons. With a geographic range stretching from the arctic to the tropics, coastal lagoons are typically found along low-lying coastlines that have a tidal range of < 4 m. Lagoons constitute 13% of coastal regions globally, range in area from < 0.01 km^2 to > 10,000 km^2, and are typically < 5 m deep.

Coastal lagoons are formed and maintained through sediment transport processes. Sediment carried by rivers, waves, currents, wind, and tides accumulates in river and tidal deltas, on marshes and flats where submerged aquatic vegetation slows currents, and on wash over fans. The process of sedimentation can eventually fill in lagoons. Lagoon barriers are constantly eroded by waves and wind, requiring continuous sediment deposition to maintain them.

Water quantity and quality in a lagoon is influenced by the rate at which the lagoon loses or gains water from evaporation, precipitation, groundwater input, surface runoff, and exchange with the ocean. Lagoon–ocean exchange is driven by tides and wave action and is often the largest component of lagoon water balance. Heat is also lost and gained through exchange with the atmosphere, sediment, and ocean.

The flushing rate, i.e., the rate at which water enters, circulates through, and exits the lagoon, is a fundamental physical property and controls the retention time of waterborne constituents. Lagoons tend to have low flushing rates because of restricted exchange with the ocean, contributing to high primary productivity and potentially high pollutant concentrations. Determinants of the flushing rate include the size and shape of the lagoon, the level of connectivity with the ocean, tidal range, and freshwater flow.

Ecological Characteristics of Lagoons

Coastal lagoons are highly productive ecosystems. They contribute to the overall productivity of coastal waters by supporting a variety of habitats, including salt marshes, seagrasses, and mangroves. They also provide essential habitat for many fish and shellfish species. For example, seagrass beds are a common feature of soft-substrate lagoons on the Atlantic coast. Where seagrass beds occur, Zostera marina (eelgrass) is the most dominant species from Maine to the Carolinas, whereas Thalassia testudinalinum (turtle grass) is the most dominant species south of the Carolinas. Such beds play an important role in influencing the shape and stability of the shoreline, regulating dissolved oxygen, and filtering suspended matter. They can enhance the biodiversity of a lagoon by providing a physical refuge from predation and also serve as nursery and feeding habitats for a variety of organisms. On the Atlantic coast, salt marshes are one of the most prevalent habitats in lagoons and are one of the most productive natural vascular plant communities in the world.

Because of their relatively low flushing rates, coastal lagoons are favorable habitats for primary producers (phytoplankton and aquatic plants). Nutrients are transported to lagoons from surface water and groundwater flows and through exchange with the ocean. Because nutrient availability often limits primary productivity, coastal lagoons can foster high rates of primary production, thereby supporting high rates of secondary production compared to other aquatic ecosystems. However, primary production that exceeds the demands of consumers can lead to eutrophication. Eutrophication is characterized by excessive phytoplankton and macroalgal blooms and subsequent hypoxia, reduced light penetration, stress and die-offs of marine organisms, loss of seagrass beds, changes in food web interactions and community structure, and loss of biodiversity.

Social Value Characteristics of Lagoons

Lagoons and their associated ecosystems are highly valued by society. However, the concept of value has different meanings in different disciplines. Among philosophers, there is debate over whether natural systems possess an intrinsic value of their own that lies outside human determination. These are complex philosophical issues that cannot be resolved here; we therefore focus only on manifestations of social values of lagoons.

Because manifestations of social values differ greatly, we organize them into four general categories: pragmatic, scholarly, inspirational, and tacit. We use the term "social values" to represent a synthesis of these categories. The placement of a manifestation of value within one category does not preclude its effect on or migration into another category. For example, although the economic engine of tourism might manifest itself as a pragmatic form of social value, the appeal of a tourist destination depends on its power to attract individuals, which in turn depends upon inspirational and tacit manifestations of social value such as dramatic scenery, pleasant swimming beaches, or shops selling mementos that capture the highlights of the tourist experience. Thus, the spectrum of four categories is intended to be a practical device for discussing specific manifestations of social value in relation to coastal lagoons.

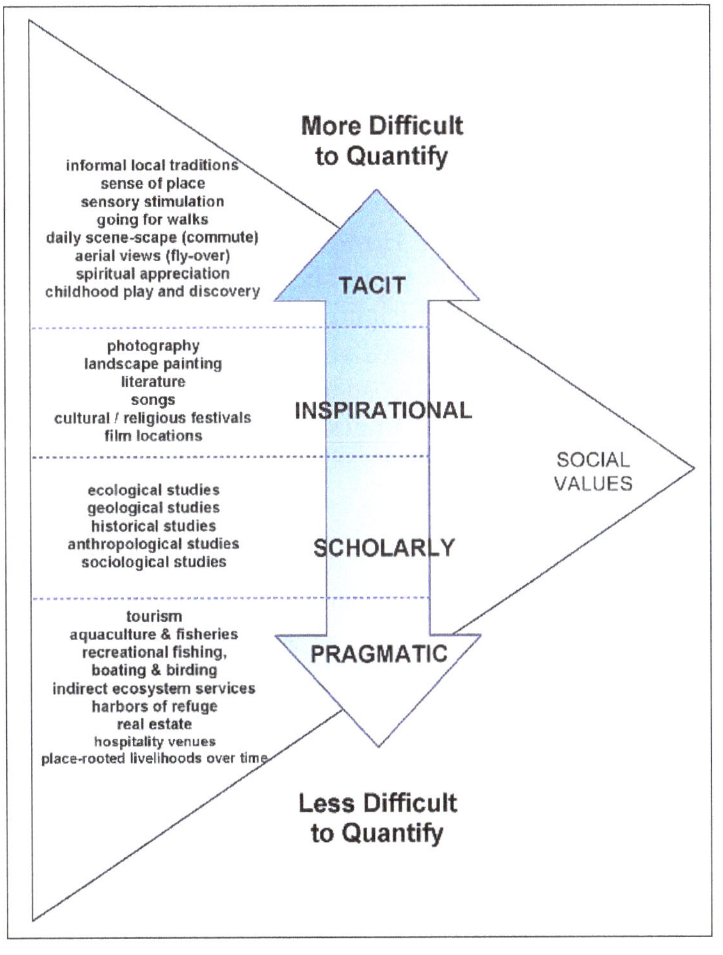

Pragmatic values are the most tangible manifestations of value and are commonly referred to as "use" values in the policy, management, and economic literature. These values reflect the commercial, recreational, and tourism uses of lagoons. Pragmatic values also include ecosystem services that indirectly support human uses. For example, salt marshes provide nursery habitat for juvenile fish that support commercial fisheries and also protect developed shorelines by reducing the impact of severe storms.

Scholarly values are embodied in activities that seek to extend our knowledge about lagoon systems and include scientific inquiry and historical study. Scholarly manifestations often contribute to pragmatic values. For example, a better understanding of lagoons can improve management, potentially increasing commercial fisheries' catch and tourism revenues. Society also values knowledge for its own sake. Thus, basic research on ecological, historical, or geological dimensions of lagoon systems provides scholarly value.

Inspirational values are ascribed to creatively productive activities that include photography and landscape painting, as well as settings for films, literature, songs, and other artistic expressions. These artistic creations encompass both personal aesthetic response to natural systems (tacit values) and profit from the sale of resulting artworks (pragmatic value).

Tacit values refer generally to unspoken values and include the enjoyment of scenery, the sounds of waves and shorebirds, and other sensory landscape features. Tacit values also encompass less tangible perceptions such as having a sense of place. In addition, society may ascribe economic existence value to an ecosystem irrespective of whether society uses it in any manner.

Some manifestations of value are more easily quantified than others, increasing the chance of their being considered and weighted more heavily in coastal management. Tacit values present the biggest challenge to quantification because they are implicit and difficult to articulate and therefore are often missing from data sets or surveys. However, they frequently manifest themselves indirectly through market transactions or other economic indicators. Small seaside cottages on stilts, for example, sit alongside lagoons or barrier islands and command million dollar price tags. Such prices reflect the powerful effect of tacit values on coastal waterfront property. The strength of tacit values is clearly evident when people choose to live in areas of high environmental risk such as low-lying coastal landscapes to benefit from aesthetic amenities or the cultural sense of place that these landscapes provide.

Inspirational values often yield a quantifiable product with high market value such as a painting, book, or film, or a popular cultural festival. Each example begins with inspirational value, but the end result can be estimated through pragmatic and scholarly means (such as tourism or cinematic criticism), thereby demonstrating the migration of inspirational values into the scholarly and pragmatic.

Social scientists have developed both qualitative and quantitative tools for incorporating all categories of value into management actions. For example, one qualitative approach developed by social scientists is the charrette, which includes unstructured discussions in public meetings. Quantitative means of measuring value, by contrast, include the revealed preference approaches developed by economists. Despite this array of tools, the central challenge remains for coastal managers to incorporate the full spectrum of values, even those that are difficult to quantify.

Table: Standard methods of quantifying values.

Method	Description	Discipline
Visual preference survey, visualization, visual impact assessment	Measures tacit manifestations of aesthetic values such as individual preferences for landscapes, architecture, or other features. Provides assessments that are primarily qualitative in nature.	Community planning, landscape architecture
Charrette	Elicits stakeholder input on a planning process or design using intensive meetings. Incorporates all categories of value into policy without providing explicit value measures. Is primarily qualitative.	Policy, planning
Revealed preference tools	Infers people's values from actual choices made, typically in a market context. Examples include hedonic pricing and travel cost methods. Provides quantitative assessments of value, typically in dollar terms.	Economics
Stated preference survey	Infers tacit manifestations of value through people's responses to hypothetical survey questions. Examples include contingent valuation and contingent choice. Provides primarily quantitative assessments of value, often in dollar terms.	Economics
Interview techniques	Ranges from informal discussion to structured surveys. Elicits values through people's responses. Provides quantitative or qualitative results, sometimes in dollar terms.	Sociology, psychology, economics, decision science
Social/civic valuation	Measures values that groups place on amenities, either through actual decisions made or through hypothetical questions posed to groups. Examples include results of public referendum and focus groups. Provides qualitative or quantitative results, sometimes in dollar terms.	Policy, planning

Table: Direct effects of projected global climate change factors on processes and physical properties of coastal lagoons.

Climate change factor				
Processes and physical properties	Sea level	Air and open ocean temperatures	Precipitation	Storms
Processes				
Barrier-lagoon migration	Erosion and shoreward migration of barrier islands			Rapid sediment redistribution
Flushing rate	Increased flushing due to barrier breaching		Variability in freshwater input increases variability in flushing	Increased flushing due to barrier breaching
Sedimentation			Sediment flux inversely related to precipitation	Increased sediment transport
Physical properties				
Lagoon water temperature		Temperature increase amplified in shallow systems		
Salinity	Change in salinity through inundation		Salinity inversely related to freshwater input	Higher salinity due to increased breaching and wash-over events
Nutrients			Higher nutrient input through overall increased terrestrial runoff	
Light penetration			Increased suspended solids during freshwater runoff events	
Oxygen		Decreased oxygen solubility with increased temperature	Lower dissolved oxygen with increased stratification	

Climate Change Impacts on Lagoon Ecosystems

Sea level, temperature, precipitation, and storminess are expected to change significantly with GCC and to impact coastal lagoons directly. Indirect impacts such as ocean acidification are important, but there is great uncertainty regarding potential effects on coastal and marine ecosystems impacts. Although both precipitation and storminess can be severe during certain weather events (e.g., hurricanes, nor'easters), we treat them as separate aspects of climate because they influence lagoons in different ways. Here, we summarize the current state of knowledge regarding these climate change variables and describe how they will likely affect the physical structure, ecological properties, and social values associated with lagoons.

Sea Level

The magnitude, rate, and effects of relative sea level increase will vary regionally because of differences in subsidence (resulting from sediment loading, groundwater withdrawal, or oil extraction), tectonic uplift, isostatic rebound, and the compaction of muddy soils. Reconstructed sea level increase estimates using tidal gauge data from 1870 to 2004 indicate that the historical average (± 95% confidence limits) rate of sea level increase was 1.7 ± 0.3 mm/yr, with significant acceleration measured at 0.013 ± 0.006 mm/yr. Satellite observations, however, show that since 1993, the sea level has been increasing at a rate of approximately 3 mm/yr, which is significantly higher than the average over the previous half-century. Global sea level increase in the 20th century measured approximately 170 mm and is expected to increase exponentially in the 21st century as air temperatures continue to increase. By 2050, the sea level is expected to have risen approximately 60–330 mm, and by 2100, is expected to have risen by 50–1400 mm above the 1990 level, with the potential to increase > 5000 mm within the next millennium.

Accelerated sea level increase is a particular threat to low-lying, shallow-gradient coastal ecosystems. Most barrier-lagoon systems respond naturally to sea level increase by migrating landward along undeveloped shorelines with gentle slopes; the retreating shore face profile can remain essentially unchanged as the shoreline retreats landward and upward in response to moderate sea level increases. However, with accelerated sea level increase, landward retreat of barriers may not be rapid enough to prevent inundation. Hardened shorelines on developed coastlines impede this natural migration and increase the vulnerability of coastal structures to inundation and storm damage.

As lagoon barriers retreat landward, accelerated sea level increase will lead to steeper and narrower barrier profiles, shortening the length of existing inlets and increasing the rate of exchange with the ocean. This will likely increase the barriers' vulnerability to breaching and the lagoons' flushing rates. As lagoons are inundated by seawater, salinity will increase, possibly altering the species composition. If a lagoon barrier persists, a sea level increase will also reduce light penetration to submerged aquatic vegetation, reducing the photosynthetic potential of these primary producers and changing the nutrient dynamics such that lagoons may be more susceptible to eutrophication.

Temperature

In the last 100 yr, global air temperature has increased by 0.76 °C and is expected to increase by an additional 1.1–6.4 °C by 2100. Temperature increases will vary regionally and will be modulated by natural climate variation. The greatest warming is expected to occur at high latitudes in winter.

Changes in air temperature strongly influence the water temperature of slow-moving, shallow water bodies such as coastal lagoons. The temperature of the world's oceans has increased, on average, by 0.3 °C and is likely to continue to increase through the next century. Because air temperatures increase more quickly over land than over oceans, coastal water temperatures are also likely to increase more rapidly. For example, in Narragansett Bay, Rhode Island, the annual average surface water temperature has increased by 1.1 °C in the past 50 yr. This rate of increase is approximately four times greater than that of the ocean. Although we are not aware of published, long-term data for water temperature in coastal lagoons, the shallow nature and low flushing rates of coastal lagoons indicate that water temperatures in lagoons will increase even more rapidly than water temperatures in open estuaries.

Water temperature in turn influences dissolved oxygen concentrations, as well as the physiology of lagoon organisms, species' ranges, and patterns of migration. Many marine species live near their threshold of thermal tolerance, at which even small changes in temperature can have large impacts on their viability. In addition, these ecosystems are more susceptible to increases in the colonization of invasive species that may thrive in warmer waters. As lagoon temperatures increase, dissolved oxygen concentrations are likely to decrease, especially as seasonal stratification of the water column isolates oxygen-depleted deeper waters from oxygen-rich surface waters. Reduced dissolved oxygen adversely affects aerobic biota, with benthic communities expected to be the most severely stressed. In lagoons with high flushing rates, the influx of ocean water will interrupt stratification, causing the water column to mix. In restricted lagoons with low flushing rates and high nutrient inputs, temperature increases will increase the probability and severity of hypoxic events. Chronic hypoxia in coastal waters has been linked to long-term changes in benthic community structure characterized by a persistent shift in species composition to more hypoxia-tolerant species and an overall decrease in species diversity.

The timing (phenology) of lagoon processes will also be affected by temperature increases. For example, changes in air temperature have affected the timing and route of migrating birds that traditionally stop over at coastal lagoons. Developmental and reproductive timing in shellfish as evidenced by early gonad maturation and spawning in bay scallops is associated with higher water temperatures. Changes in plankton phenology have also been observed, with zooplankton becoming active earlier in the year following relatively warm winters. Increased grazing of phytoplankton by zooplankton has been observed and may partially explain declines in phytoplankton abundance associated with warmer temperatures. Warmer temperatures are also thought to contribute to observed declines in seagrass abundance.

Precipitation

There is general agreement that there will be changes in precipitation intensity, timing, volume, and form (snow or rain) because of GCC. Estimates of the magnitude and timing of changes in precipitation patterns along the U.S. Atlantic and Gulf coasts are variable, reflecting the high level of uncertainty associated with these predictions. Although predictions agree that there will be an increase in extreme rainfall events and drought–flood cycles, disagreement persists regarding the direction of expected changes in other components of the water cycle such as stream flow (timing and volume), evapotranspiration, soil moisture, and groundwater storage resulting from complex interactions between and among factors such as air temperature and seasonal timing of precipitation events.

Regional changes in precipitation patterns can have important effects on the physical and ecological characteristics of coastal lagoons through the alteration of freshwater inputs and associated changes in salinity and dissolved oxygen concentrations. Intense precipitation events would increase short-term freshwater inputs while locally decreasing salinity. Conversely, lower precipitation would reduce freshwater inputs and potentially result in higher salinity. Salinity would also be affected by changes in flushing rate, which may counteract the changes in freshwater inputs. Increased freshwater inputs can also contribute to the stratification of deeper lagoons, increasing the risk of hypoxia in bottom waters. The expected increase in the variability and intensity of precipitation events is therefore expected to produce increased variability, both spatially and temporally, in salinity and dissolved oxygen concentrations in coastal lagoons.

Other effects of increased surface water inputs include the increased delivery of sediment and nutrients to lagoons. Increased nutrient inputs may accelerate the eutrophication of lagoons, especially those with low flushing rates. As with sea level increases, increased turbidity will reduce light penetration and the photosynthetic activity of submerged aquatic vegetation, compounding the risk of eutrophication as nutrient dynamics are further altered. In addition, reduced light penetration can inhibit the feeding ability of visual predators.

Storms

Storms affect lagoons through over wash events and by erosion from wind and waves. There is disagreement among predictions concerning the extent and magnitude of changes in storminess with GCC. Evidence of an increase in intense tropical cyclone activity in the North Atlantic over the past 40 yr supports predictions that the frequency and intensity of extreme weather events have been increasing and will continue to increase with warmer global temperatures. However, these predictions have been challenged by suggestions that the apparent trend in increasing storm frequency is an artifact of improved monitoring and by predictions that increased vertical wind shear could dampen the effects of increasing hurricane intensity. Meehl et al. suggest that a warmer climate will increase the overall intensity of tropical cyclones and, whereas the number of storms is expected to decrease globally by the end of the 21st century, the number of storms in the North Atlantic could increase by as much as 34% during this period.

The consequences of sea level increase become acute during storm events because sea level increase and storms interact to erode barriers and produce high storm surges, rapidly redistributing barrier sediment. During periods of high storm surge, water moves rapidly over the barrier in a process called overwash which delivers sediment eroded from the front of the barrier onto the back barrier flat and into the lagoon. Increased storm intensity will likely cause more frequent breaches of barrier islands, which in turn will increase the rate of exchange with the ocean and consequently the flushing rate and salinity of lagoons.

Climate Change Impacts on Social Values of Lagoons

The combined effects of increased temperature, sea level, and storminess and the variability in weather patterns will likely stress human communities that value the presence of particular lagoons. Lagoons are valued for many reasons: as ecosystems that support an abundance of species on which livelihoods depend (pragmatic), as sites of historical tradition (scholarly), as incentive for creative pursuits (inspirational), and as a source of a sense of place (tacit).

When faced with GCC stressors, society may take actions to reduce impacts. However, these actions may also have unintended consequences, creating a feedback cycle that can have long-term negative consequences. For example, people have historically reacted to coastal erosion by hardening shorelines with sea walls or riprap. Hardened shorelines may protect coastal structures in the short term, but they can exacerbate erosion elsewhere by reducing sources of sand to replenish shorelines downcoast.

Table: Direct connections between social, global climate change (GCC), and ecological impacts in four coastal lagoons.

Lagoon system				
Characteristic	Barataria Bay	Rhode Island salt ponds	Carolina Outer Banks and Lowcounty	Jamaica Bay
Geography/social	Barataria Bay lagoon systems: these largely rural settings are replete with small settlements populated by marsh dwellers who live as part of the landscape.	A "string of pearls" of shallow coastal salt ponds representing a range of lagoon settings: tourism sprawl, exurban, and nature refuges.	A strand of islands and barrier beaches rich in history from the traditionally accepted founding of the first colony in America to the influence of the slave trade and the Civil War. These barrier islands also have a history of extreme vulnerability to hurricanes.	A lagoon located within Gateway National Park in the outer boroughs of New York City (specifically, Brooklyn and Queens), directly adjacent to John F. Kennedy International Airport. The park attracts 6 million visitors each year.
Anticipated GCC impacts	Air temperature: Predicted to increase by 1.1–2.2°C by 2100. Precipitation: Increase of 10% in summer and fall11 leads to 1–3 parts per trillion decrease in salinity. Sea level: Rate of increase is faster than marsh accretion rates. Storms: Vulnerable to tropical storms and hurricanes, which cause devastation of marsh grasses, wash-over fans, and erosion from shallow areas.	Air temperature: Predicted to increase by 1.9–6.9°C by 2100. Sea level: Accelerated increase will increase erosion of barrier beaches, landward migration of barriers, and extensive loss of coastal wetlands. Storms: High degree of shoreline change due to the narrow and low profile of barrier beaches. Sea level: The relative increase is between 3 and 4 mm/yr.	Precipitation: High flow to estuaries with large watersheds are most common. Storms: Vulnerable to tropical storms and hurricanes, which accelerate inlet formation, wash-over fans, and inundation.	Storms: More frequent storm surges and greater wave action superimposed on sea level increase will exacerbate marsh erosion.
Ecological impacts (examples)	Increased salinity from sea level increase will result in the loss of nursery and mating habitat, affecting juvenile penaeid shrimp and blue crabs, both of which are commercially important.26,27 Conversely, the increased marsh subsidence will benefit shrimp by increasing the amount of time the marsh surface is available.	Temperature stress and increased water depth and salinity from sea level increase will affect seagrass habitats.29 Shifts in species, species distribution, and abundance of seagrasses may occur, affecting local species that are dependent on seagrass habitat such as juvenile flounder.	Migratory birds that depend on barrier beaches will experience the loss of habitat.30 Low oxygen conditions resulting from increased eutrophication will have negative impacts on vulnerable species such as striped bass.	Without adequate sediment supply and with increased erosion, populations of horseshoe crabs will be affected by the loss of suitable shoreline for breeding habitat.31 Decreases in food supply from decreases in horseshoe crab eggs and in salt marsh habitat will negatively affect migrating shorebirds.

To explore how particular social structures and physical and ecological characteristics are likely to be impacted by GCC in the coming decades, we examined four specific lagoon systems on the Atlantic and Gulf coasts: the Gulf Coast of Louisiana, the salt ponds of Rhode Island, the Outer Banks and Lowcountry of the Carolinas, and Jamaica Bay, New York. Table provides a brief summary of the socio-geographic settings of the four lagoon systems, as well as key GCC factors and some likely ecosystem impacts. These lagoons were selected to cover a broad geographic range and include systems with distinct human development patterns and a variety of social values.

Gulf Coast of Louisiana

Coastal Louisiana is a vast rural complex of lagoons, bayous, canals, channels, and small embayments that are particularly vulnerable to sea level increase. These coastal lands are disappearing at a rate of up to 100 km^2/yr because of the combined effects of sea level increase, hurricanes, subsidence and erosion. Without abatement, an additional 3200 km^2 of wetlands are predicted to be inundated by 2040, advancing the Louisiana shoreline inland by as much as 53 km in some areas. These losses are caused in part by human actions in response to flooding threats, including increased channelization of the Mississippi River with levees, which has controlled flooding, but has also cut off sediment supply to delta wetlands and has diverted sediments to deep waters off the continental shelf. This has exacerbated the effects of sea level increase with significant negative impacts on barrier islands and wetlands.

Coastal Louisiana's lagoons and wetlands are spawning, nursery, and fishing grounds for many important fisheries. For example, shrimp landings in Louisiana constitute roughly 30% of the ex-vessel value of the national shrimp catch: Sea level increase will intensify the loss of critical salt marsh habitat. Flooding from sea level increase will increase salinity and may stress juvenile shrimp that mature in low-salinity estuarine waters. The shrimp fishery has significant pragmatic value in this region and is also a unique way of life that embodies inspirational and tacit values.

Louisiana's Barataria Bay sustains rich biodiversity by providing not only important habitat for waterfowl, shrimp, and finfish, but also serving as homeland to several cultures, creating a unique sociological phenomenon in which highly diverse communities (Houma, Cajun, Creole,

African-American, Vietnamese, and Canary Island Spaniards) share a single ecosystem. These cultures share a high degree of poverty or economic uncertainty, leaving them more vulnerable to the effects of GCC.

Moreover, the inundation of Louisiana's coastal lagoons increases the vulnerability of New Orleans to flooding, risking the loss of one of the world's iconic cities and the living culture that produced jazz, one of America's indigenous art forms. With the loss of these coastal lagoons, a powerful experience of place is also lost, existing only in nostalgic representations preserved in artworks such as those of Colette Pope Heldner and John James Audubon or the writings of James Lee Burke.

Salt Ponds of Rhode Island

The string of lagoons along the south shore of Rhode Island are known locally as salt ponds and figure prominently in a complex landscape of highly developed residential areas, commercial waterfronts, agricultural areas, and uninhabited protected refuges. These lagoons have historically played an important role in Native American culture, and contemporary human uses include commercial and recreational fishing, boating, birding, surfing, and photography. Local communities depend upon the fishing industry, and these lagoons provide habitat for a variety of commercially important shellfish and finfish.

Rhode Island's salt ponds are currently under heavy pressure from residential development. Past construction of stabilized breach ways dramatically altered the salinity in many of these lagoons, shifting the lagoon community from brackish species (alewives, smelt, oysters) to species that thrive in more saline conditions and are viewed as pragmatically valuable to local communities. Sea level increase will likely increase the lagoons' salinity through breaching of barriers, promoting another shift in species composition. Temperature increases and reduced water clarity will be detrimental to eelgrass beds and will impede restoration efforts to increase the areal extent of these submerged aquatic vegetation habitats. Storminess and sea level increase are of particular concern because hard-edged coastal development (i.e., permanent structures) restricts the migration of barrier and headland shorelines. Pragmatic values will be impacted through property loss as shorelines continue to shift because of sea level increase and increased storm intensity. Impacts are compounded because real estate on lagoons is some of the most expensive property in these coastal communities. The means for the continued survival of the salt ponds is limited by attempts to protect structures through the hardening of lagoon shorelines, yet another example of feedback between social and ecological systems. The potential loss of Rhode

Island's salt ponds also threatens the sense of place that inspires significant artistic enterprise, e.g., landscape painting (John Lutes), cartooning (Don Bosquet), and literature, as well as birding and the seasonal pilgrimage of tourists to beaches along barriers.

Outer Banks and Lowcountry of the Carolinas

The Carolina Outer Banks are barrier spits that separate a large estuarine system, including Pamlico Sound, the largest Atlantic coastal lagoon, from the Atlantic Ocean. The Carolina Lowcountry consists of a series of lagoons and waterways that extend along the South Carolina coast. Both systems provide habitat for migratory birds, halophilic grasses, finfish, and shellfish. The Outer Banks and Lowcountry are a transitional cultural setting in which the wealthy compete for coastal land with poorer communities of longstanding historical importance. Islands off the South Carolina coast have been home to generations of African Americans in a unique sea-island culture known as the Gullah in South Carolina and Geechee in Georgia. Gullah-populated islands, including Edisto Island, Coosaw Island, St. Helena Island, and Daufuskie Island, served as the inspiration for the fictional Yamacraw Island in Pat Conroy's The Water is Wide. The Gullah culture inspired Julia Peterkin's Pulitzer Prize winning novel Scarlet Sister Mary and the paintings of John W. Jones. These works exemplify scholarly and inspirational values.

Sea level increase threatens the low-elevation barriers in this region and will change the hydrology and salinity of the lagoons. The catchment area draining into this lagoon system is relatively large compared to the area of the lagoons, intensifying the effect of increased precipitation. Nutrients carried with higher freshwater flows increase the likelihood of eutrophication, affecting not only pragmatic values associated with fisheries and real estate, but also inspirational and tacit values. These coastal lagoons, including traditional Gullah lands, are threatened by residential and commercial development, which will in turn be threatened by inundation from hurricanes and sea level increase. Residents of low socioeconomic status may not have the capacity to evacuate flood-prone areas during extreme storm events, as was the case with many New Orleans communities during Hurricane Katrina in 2005. Under current climate change projections, the delicate balance between local cultural groups and their lagoon landscapes will be severely disrupted, with the potential for permanent displacement of a singular indigenous culture and language unique to the Gullah. A GCC-induced diaspora will diminish the related scholarly, inspirational, and tacit values of these lagoons. In addition, pragmatic and scholarly values are at risk from erosion, sea level increase, hurricanes, and storm surge that threaten significant archaeological sites and other historic landmarks such as the Cape Hatteras lighthouse. Contrary to the traditional strategy of responding to coastal erosion by hardening shorelines, protection of the Cape Hatteras lighthouse was achieved through relocation in 1999.

Jamaica Bay, New York

Jamaica Bay is a coastal lagoon located in one of the most densely developed urban settings in the United States and represents the largest natural open space in New York City. As part of the Gateway National Recreation Area (U.S. National Park Service), Jamaica Bay hosts > 6 million visitors annually and provides habitat for > 300 species of migratory and resident birds, as well as spawning grounds for horseshoe crabs. The lagoon provides urban residents direct access to a relatively undeveloped natural world and is a living classroom for schoolchildren involved in National Park Service sponsored curricula that link classroom science activities with natural ecosystems.

The salt marshes of Jamaica Bay that provide prime habitat have been receding at an accelerating pace. Although activities such as channel dredging and heavy boat traffic have likely contributed to short-term marsh loss, encroaching sea level increase and increased storminess present long-term hazards to Jamaica Bay's lagoon system. As GCC impacts unfold, urban residents could lose the tacit benefits of this contrasting land and lagoon scape and the calming effects of ambient natural sights, sounds, and textures.

Coral Reefs

Coral reefs are some of the most diverse ecosystems in the world. Coral polyps, the animals primarily responsible for building reefs, can take many forms: large reef building colonies, graceful flowing fans, and even small, solitary organisms. Thousands of species of corals have been discovered; some live in warm, shallow, tropical seas and others in the cold, dark depths of the ocean.

Because of the diversity of life found in the habitats created by corals, reefs are often called the "rainforests of the sea." About 25% of the ocean's fish depend on healthy coral reefs. Fishes and other organism shelter, find food, reproduce, and rear their young in the many nooks and crannies formed by corals. The Northwest Hawaiian Island coral reefs, which are part of the Papahānaumokuākea National Marine Monument, provide an example of the diversity of life associated with shallow-water reef ecosystems. This area supports more than 7,000 species of fishes, invertebrates, plants, sea turtles, birds, and marine mammals. Deep water reefs or mounds are less well known, but also support a wide array of sea life in a comparatively barren world.

Coral Characteristics

Shallow water, reef-building corals have a symbiotic relationship with photosynthetic algae called zooxanthellae, which live in their tissues. The coral provides a protected environment and the compounds zooxanthellae need for photosynthesis. In return, the algae produce carbohydrates that the coral uses for food, as well as oxygen. The algae also help the coral remove waste. Since both partners benefit from association, this type of symbiosis is called mutualism.

Deep-sea corals live in much deeper or colder oceanic waters and lack zooxanthellae. Unlike their shallow water relatives, which rely heavily on photosynthesis to produce food, deep sea corals take in plankton and organic matter for much of their energy needs.

Benefits of Coral Reef Ecosystems

Coral reefs protect coastlines from storms and erosion, provide jobs for local communities, and offer opportunities for recreation. They are also are a source of food and new medicines. Over half a billion people depend on reefs for food, income, and protection. Fishing, diving, and snorkeling on and near reefs add hundreds of millions of dollars to local businesses. The net economic value of the world's coral reefs is estimated to be nearly tens of billionsoffsite link of U.S. dollars per year. These ecosystems are culturally important to indigenous people around the world.

Threats to Coral Reef Ecosystems

Unfortunately, coral reef ecosystems are severely threatened. Some threats offsite link are natural, such as diseases, predators, and storms. Other threats are caused by people, including pollution, sedimentation, unsustainable fishing practices, and climate change, which is raising ocean temperatures and causing ocean acidification. Many of these threats can stress corals, leading to coral bleaching and possible death, while others cause physical damage to these delicate ecosystems. During the 2014-2017 coral bleaching event, unusually warm waters (partially associated with a strong El Niño) affected 70% of coral reef ecosystems worldwide. Some areas were hit particularly hard, like the Great Barrier Reef in Australia, where hundreds of miles of coral were bleached.

Corals are able to recover from bleaching events if conditions improve before they die, though it can take many years for the ecosystems to fully heal. Scientists are also testing new ways to help coral reef ecosystems, such as growing coral in a nursery and then transplanting it to damaged areas.

References

- Marine-ecology, conservation: marinebio.org, Retrieved 15 January, 2019
- Xiao-feng, pang (2014) water: molecular structure and properties, chapter 5, pp. 390–461, world scientific. Isbn 9789814440448
- Marine-life-definition-and-examples-2291890: thoughtco.com, Retrieved 1 June , 2019
- Martindale, mark; finnerty, j.r.; henry, j.q. (september 2002). "the radiata and the evolutionary origins of the bilaterian body plan". Molecular phylogenetics and evolution. 24 (3): 358–365. Doi:10.1016/s1055-7903(02)00208-7.
- Marine-biology: biologydictionary.net, Retrieved 1 August , 2019
- Threats-to-marine-environment-you-must-know, environment: marineinsight.com, Retrieved 9 May, 2019
- Rocky-shore-habitat, wiki: coastalwiki.org, Retrieved 8 August , 2019
- Estuary, science: britannica.com, Retrieved 13 July, 2019
- Coral-reef-ecosystems, marine-life-education-resources, resource-collections, education: noaa.gov, Retrieved 3 June, 2019

3

River Ecology

The branch of ecology which deals with the study of flowing waters as well as the various abiotic and biotic interactions among animals, plants and microorganisms within rivers is termed as river ecology. The diverse aspects of river ecology such as the restorations of rivers have been thoroughly discussed in this chapter.

The ecology of the river refers to the relationships that living organisms have with each other and with their environment – the ecosystem. An ecosystem is the sum of interactions between plants, animals and microorganisms and between them and non-living physical and chemical components in a particular natural environment.

River ecosystems have:

- Flowing water that is mostly unidirectional.
- A state of continuous physical change.
- Many different (and changing) microhabitats.
- Variability in the flow rates of water.
- Plants and animals that have adapted to live within water flow conditions.

Kōtare or kingfishers that inhabit river areas eat small fish, insects and freshwater crayfish. If these river species decline, the kingfisher will move to another habitat.

Water Flow

Water flow is the main factor that makes river ecology different from other water ecosystems. This is known as a lotic (flowing water) system. The strength of water flow varies from torrential rapids to slow backwaters. The speed of water also varies and is subject to chaotic turbulence. Flow can be affected by sudden water input from snowmelt, rain and groundwater. Water flow can alter the shape of riverbeds through erosion and sedimentation, creating a variety of changing habitats.

The Waikato River is a lotic system. Lotic means flowing water.

Substrate

The substrate is the surface on which the river organisms live. It may be inorganic, consisting of geological material from the catchment area such as boulders, pebbles, gravel, sand or silt, or it may be organic, including fine particles, leaves, wood, moss and plants. Substrate is generally not permanent and is subject to large changes during flooding events.

Light

Light provides energy for photosynthesis, which produces the primary food source for the river. It also provides refuges for prey species in the shadows it casts. The amount of light received in a flowing waterway is variable, for example, depending on whether it's a stream within a forest shaded by overhanging trees or a wide exposed river where the Sun has open access to its surface. Deep rivers tend to be more turbulent, and particles in the water increasingly weaken light penetration as depth increases.

Kaitiaki are working to restore and protect the health and wellbeing of the Waikato River.

Temperature

Water temperature in rivers varies with the environment. Water can be heated or cooled through radiation at the surface and conduction to or from the air and surrounding substrate. Temperature differences can be significant between the surface and the bottom of deep, slow-moving rivers.

Climate, shading and elevation all affect water temperature. Species living in these environments are called poikilotherms – their internal temperature varies to suit their environmental conditions.

Water Chemistry

The chemistry of the water varies from one river ecosystem to another. It is often determined by inputs from the surrounding environment or catchment area but can also be influenced by rain and the addition of pollution from human sources.

Oxygen is the most important chemical constituent of river systems – most organisms need it for survival. It enters the water mostly at the surface, but its solubility decreases as the water temperature increases. Fast, turbulent waters expose a wider water surface to the air and tend to have lower temperatures – achieving more oxygen input than slow backwaters. Oxygen is limited if water circulation is poor, animal activity is high or if there is a large amount of organic decay in the waterway.

Bacteria

Bacteria are present in large numbers in river waters. They play a significant role in energy recycling. Bacteria decompose organic material into inorganic compounds that can be used by plants and by other microbes.

Plants

Plants photosynthesise – converting light energy from the Sun into chemical energy that can be used to fuel organisms' activities.

Underwater plants

A variety of plants can be found growing within a river system. Some plants are free-floating while others are rooted in areas of reduced current.

Algae are the most significant source of primary food in most rivers or streams. Most float freely and are therefore unable to maintain large populations in fast-flowing water. They build up large numbers in slow-moving rivers or backwaters. Some algae species attach themselves to objects to avoid being washed away.

Plants are most successful in slower currents. Some plants such as mosses attach themselves to solid objects. Some plants are free-floating such as duckweed or water hyacinth. Others are rooted in areas of reduced current where sediment is found. Water currents provide oxygen and nutrients for plants. Plants protect animals from the current and predators and provide a food source.

Invertebrates

Invertebrates have no backbone or spinal column and include crayfish, snails, limpets, clams and mussels found in rivers. A large number of the invertebrates in river systems are insects. They can be found in almost every available habitat – on the water surface, on and under stones, in or below the substrate or adrift in the current. Some avoid high currents by living in the substrate area, while others have adapted by living on the sheltered downstream side of rocks. Invertebrates rely on the current to bring them food and oxygen. They are both consumers and prey in river systems.

Fish

The ability of fish to live in a river system depends on their speed and duration of that speed – it takes enormous energy to swim against a current. This ability varies and is related to the area of habitat the fish may occupy in the river. Most fish tend to remain close to the bottom, the banks or behind obstacles, swimming in the current only to feed or change location. Some species never go into the current. Most river systems are typically connected to other lotic systems (springs, wetlands, waterways, streams, oceans), and many fish have life cycles that require stages in other systems. Eels, for example, move between freshwater and saltwater. Fish are important consumers and prey species.

Banded kōkopu are found in the Waikato River system.
They are one of the species that make up whitebait.

Birds

A large number of birds also inhabit river ecosystems, but they are not tied to the water as fish are and spend some of their time in terrestrial habitats. Fish and water invertebrates are an important food source for water birds.

River at Length

Longitudinal Structure of the River Corridor

All rivers, regardless of their type, have the same stages of structural changes from source to delta.

The profile of every river can be divided into three zones.

- Source zone (headwaters) – Mountain streams that slope rapidly downhill and form V-shaped river valleys, often forming waterfalls. These rapids shape and carry large-size sediments downstream.
- Transfer zone – Characterized by a lower altitude. The flow velocity is slower, the river bed becomes wider, and meanders form. Some of the larger sediments settle at the interim area between the source zone and the transfer zone, thus forming the so-called sediment cones. Other sediments are carried further downstream. Erosion and deposition processes are in equilibrium in the transfer zone. Most of the lowland water courses in Bulgaria belong to this zone, including the Bulgarian section of the Danube River.
- Deposition zone – Of rather low slope; flow velocities are slow, forming wide meanders. Most of the sediments, including the finest ones, settle in this area. The river mouth often opens into a wide delta bottomed by fine sediments, and the river splits into many arms. In Bulgaria such zones are formed only in the lower course of rivers that flow directly into the Black Sea.

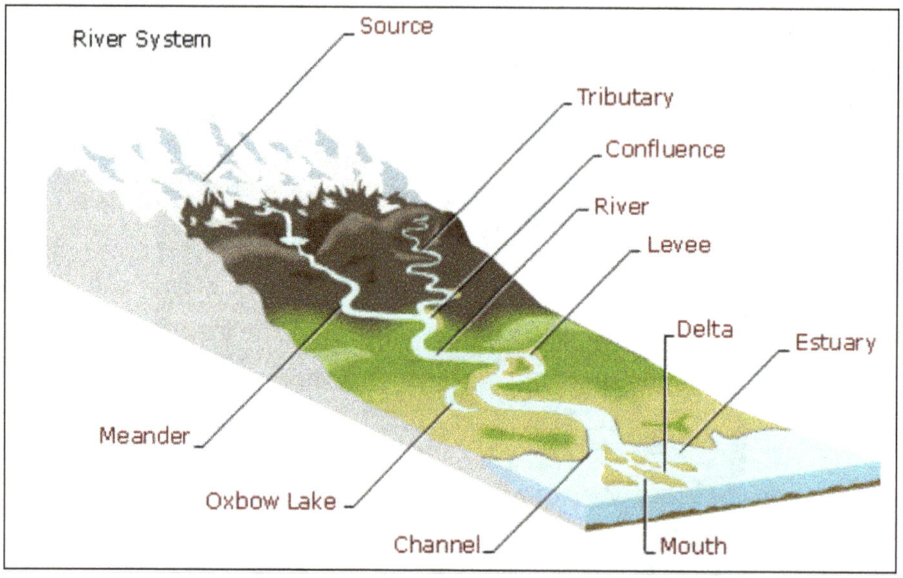

Length-wise zoning of a river.

Sinuosity

Natural rivers are never straight. We can calculate the sinuosity index by dividing the mid-river length by the length of the river valley. When the sinuosity index is more than 1.3, the river is defined as meandering. This index is invariable for the relevant river section. This means that if the river course becomes straight as a consequence of human interference or natural "shortcutting" of a meander, disbalance will occur and the river will try to restore its previous sinuosity.

Such successive sequencing is also a precondition for quick and complete dissolution of pollutants. It is not by chance that straightened (regulated) rivers with altered sequencing of pools and riffles lose not only a significant part of their fish population and other aquatic life, but also most of their self-purification capacity.

Riffles and Pools

Regardless of the form of their beds, all rivers have equidistant sequencing of shoals and deeper sections, called riffles and pools. They are linked to the river midstream – a line which benchmarks the deepest section of the river and where the water flows fastest. Pools are formed in the midstream, next to the outside bank of a watercourse bend. Riffles are formed between two bends where the midstream goes from one river bank to the other. The distance between two pools or two riffles – a common feature for all rivers – is 5 to 7 times the width of the river, measured at its highest water level.

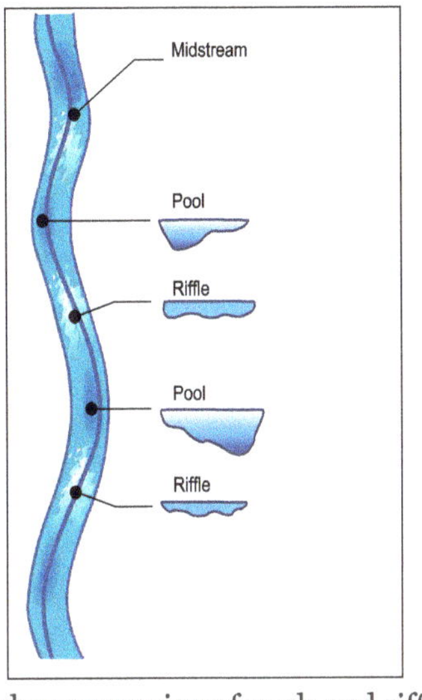

Regular sequencing of pools and riffles.

The sequencing of pools and riffles within a river section, and the existence of various habitats illustrates the variety of life adapted to both fast and slow currents and serves as a precondition for a wellfunctioning river ecosystem.

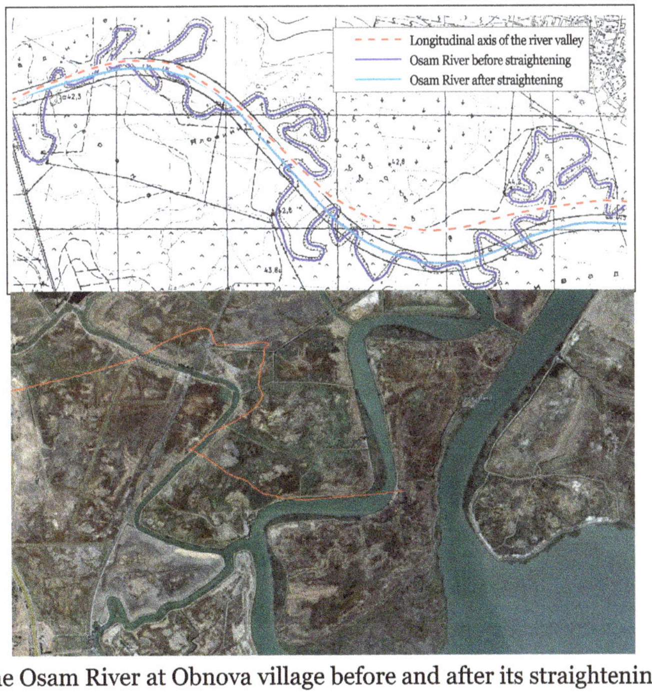

The Osam River at Obnova village before and after its straightening.

Before the straightening of the Osam River, the sinuosity index in the illustrated section was 3.55, which defined it as strongly meandering. After the straightening, the index became 1.02 which defines the river as "straight". The shortcut of the river course reduced its length more than 3 times its natural length.

River at Width

Horizontal Structure of the River Corridor

The basic component of river corridor is the riverbed. This is a naturally lowered landform usually used as a passageway for flowing water. The next component is the floodplain – this is the neighbouring part of the riverbed, flooded by high water at different intervals varying from very often to seldom. The floodplain may be situated on either one or both river banks, depending on relief. The next component is the transition upland fringe, which begins above the floodplain and forms the frontier between the river corridor and the adjacent landscape.

Riverbed (Bottom)

The shape and size of the riverbed are defined by the equilibrium of four basic factors – the energy of the water stream (slope and water velocity) and the resistance of substratum (particle size of the sediment and its runoff). If we illustrate this on a scale where we have the sediments on one side and the water energy on the other, every decrease in the water energy will result in smaller sediment particles and in the widening of the riverbed. The river equilibrium is reached when all four variables are balanced.

Floodplain

The floodplain is theoretically divided into two parts: hydrologic floodplain and topographic floodplain. The hydrologic floodplain is an area inundated about two out of every three years. The topographic floodplain includes the hydrologic be developed in the floodplains.

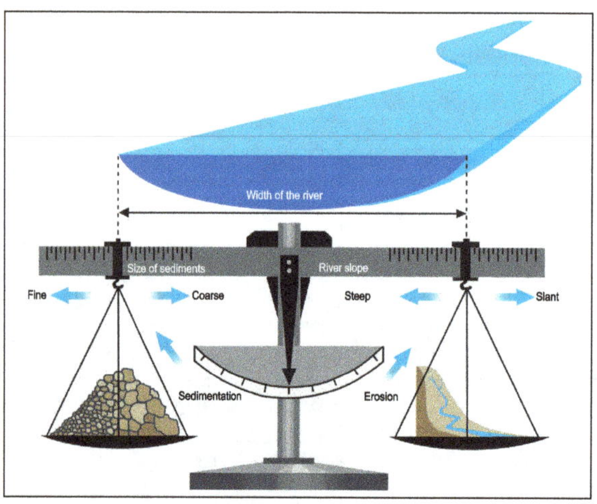

The hydrologic floodplain is an area inundated about two out of every three years. The topographic floodplain includes the hydrologic floodplain up to the altitude reached by a flood peak of given frequency (every 100 or 500 years). The topographic floodplain is used in spatial planning and development, and all the activities therein must consider the risk of flooding. For instance, no human settlements or important infrastructure facilities should be developed in the floodplains.

The floodplains, especially those downstream, are dynamic systems with many various components that undergo a continuous process of formation or disappearance:

- Meander scrolls – Old alluvial deposits that mark the former passageway of the riverbed.
- Oxbow lakes or abandoned river bends – Former river meanders that were gradually cut off the riverbed. They can be in different stages of development – from those still linked to the river to those which are rather far away and almost dried up.
- Natural levees or sand-bars – These are formed during flooding when large amounts of sediments are carried out by waters that disperse, losing most of their energy and depositing the coarser sediments on one of the river banks, thus forming an embankment.
- Splay – Structures that form at high water when streams under pressure break through natural levees and spread deposits into a fan-shaped structure.
- Backswamp – Wetlands that are formed after high water as a result of natural levee formation which prevents water going back to the riverbed, thus prolonging the flooding period.

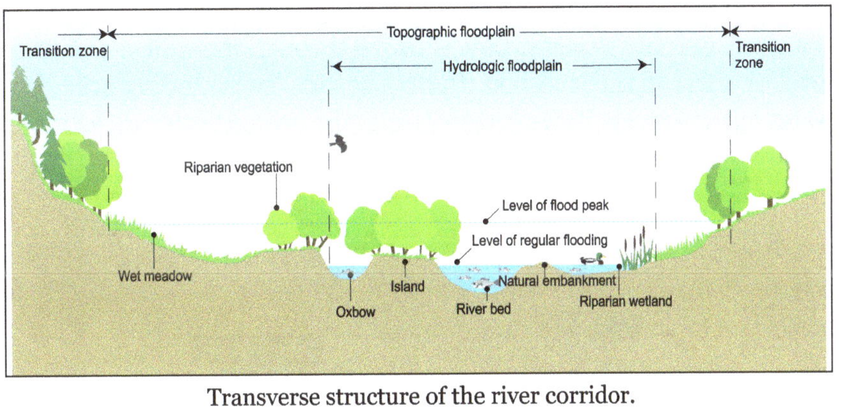

Transverse structure of the river corridor.

Wildlife in rivers depends on all these floodplain components. For instance, many of the carp species need backswamps and oxbow lakes to spawn. These are also the places where fish fry find food and shelter amongst the water vegetation. Draining of riparian wetlands or cutting off their link with the river is one of the main reasons that downstream fish population decreases.

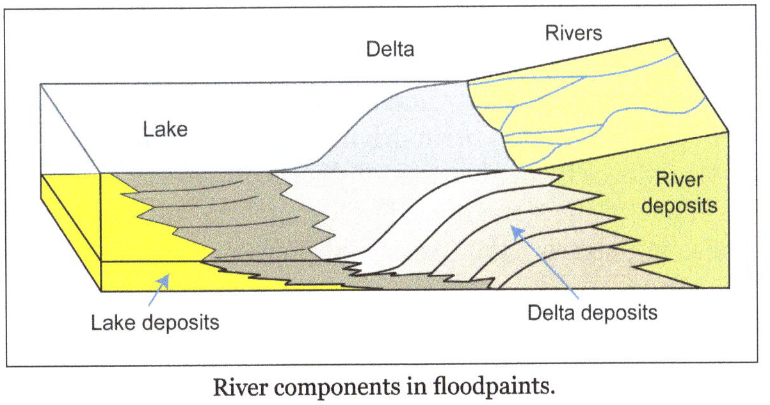

River components in floodpaints.

Rivers in Depth

Vertical Structure of the River Corridor

The water level of most rivers varies greatly depending on the season. The seasonal level of rivers usually varies between 10-20 cm and 6-7 m (such as the Danube). The seasonal increase of water level has an important role in supporting river ecosystems. Rivers often link to adjacent wetlands only at high water level, which is often the only possibility for fish populations to migrate.

Unique vegetal systems develop on the banks of rivers with higher variations of water levels, especially in the lower course; they are very well adapted to such conditions. These are the floodplain forests. One characteristic feature of their vegetation is the impenetrable lianas.

Due to their appearance and extremely rich variety of animals and plants, the floodplain forests in Bulgaria are reminiscent of an equatorial jungle. In the past, floodplain forests were widespread in Bulgaria. Today such forests are only partly preserved along Danube islands, the rivers of Kamchiya, Ropotamo, Tundja, and the islands of the Maritsa River; on the other hand such floodplain forests have been completely destroyed along the Iskar, Vit, Osam, Yantra and Sturma rivers.

River depth does not end where its bottom lie. The riverbed usually lies on water-saturated, old river sediments. This groundwater runs to or from the river down the river valley. The stream in the riverbed is usually just a small part of the water flow running through the valley. The level of groundwater depends on the level of water in the riverbed. In dry seasons, surrounding areas drain into the river bed and groundwater runs towards it. During flooding, the groundwater direction may reverse.

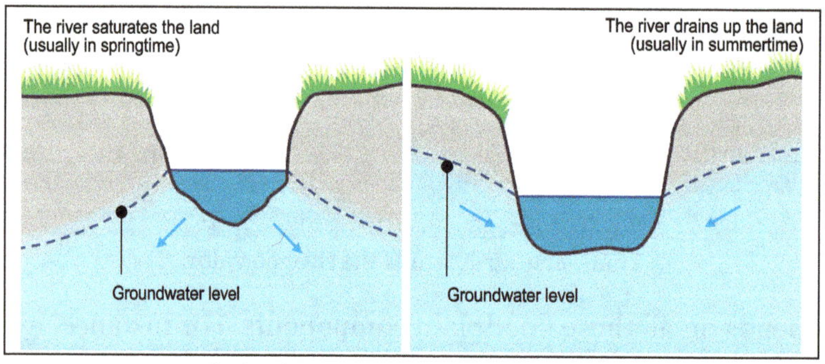

Relation between river and groundwater.

River in Time

River Runoff and Flooding

Water transfer from precipitation into rivers and oceans and vice versa, including its vaporization, is known as the "water cycle". This is the main driving force of the river ecosystem. River runoff is the quantity of water that moves through a spot of the river over certain time. This indicator can vary within very wide limits for the same river. For instance, in some torrential catchment basins (almost half of the basins in Bulgaria), the runoff may increase 100 times after a downpour.

The seasonal increase of water runoff and associated flooding are natural and unavoidable processes. With regard to people, flooding may cause serious consequences – from positive (the ancient Egypt civilization developed on the banks of the Nile River) to disastrous ones.

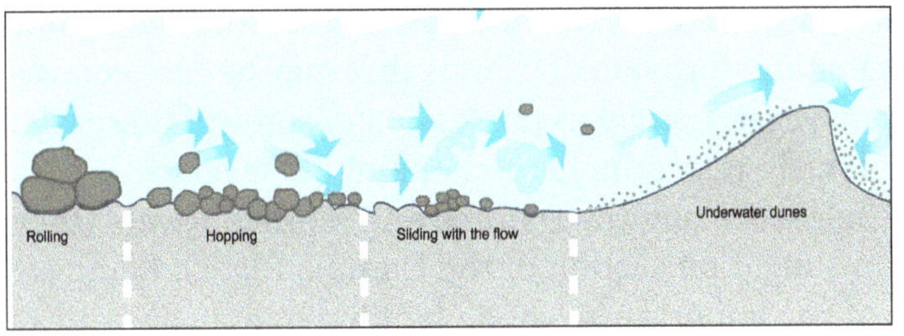

Forms of sediment movement.

Flooding does not "damage" nature. On the contrary, this is a cyclic process that serves an important role for the normal functioning of the river corridor. Spring high water is a prerequisite for fish migration.

The regular water flow through riparian wetlands is also important because it carries away dead organic material and cleans the area. Many back swamps exist due to periodical flooding. If this process and the link with the rivers are broken, these areas will silt up, the succession or "natural aging" will accelerate and they will finally disappear.

Sediment Transport

The movement of eroded soil and rock particles in water flow is called transport of sediments. Sediment transport is measured by the number of undissolved particles that pass through a river spot over a certain time. The energy required to carry the particles and move them downstream relates to the difference between the velocities of water in different strata.

Water in bottom layers moves more slowly due to higher friction, while water in upper layers flows faster due to lack of friction. Thus a whirlpool, or vortex, is formed between the two water layers – this spiraling bottom. They roll, hop or slide in the water depending on their size and density.

The sediment load of a river consists of suspended and bedload sediment:

- Suspended sediments are fine particles which are easy to move by turbulence in the bottom layer. These particles can be taken up high in the water column and transported a long distance with no further contact with the riverbed as long as the water has enough turbulence. The main part of the river solid load (50-90%) consists of suspended sediments. The source of suspended sediments is the entire river basin and the fine particles brought by rainwater and as a result of the soil erosion on the river banks.
- Bedload sediments are the bottom particles, which are dense and big enough to move downstream mainly by rolling or hopping. The size of bedload sediment is the same as that in the river bed in the given river section. The source of bedload sediments is the riverbed itself and, to a much lesser extent, the erosion of the banks and other sources.

Riverbed Processes

The processes of erosion, sediment transport, and deposition interacting with vegetation make river corridors very dynamic ecosystems. Rivers constantly change shape and location. New meanders and arms are formed, old ones are abandoned; the riverbed itself shifts. This phenomenon is called migration of the river corridor.

Sand dunes in the lower part of the river course also shift from their initial location. Sometimes they surface over the water and form small islands that may be carried away and destroyed by the next high floods. Riparian vegetation takes possession of some of these islands which can exist for several decades, continuously relocating and changing their shape Figure. We can find most of the riparian islands in Bulgaria along the Danube and Maritsa Rivers (there are about 120 to 140 islands in the Romanian-Bulgarian section of the Danube River). Being aware of and monitoring riverbed processes are of key importance when planning human activities in river valleys – especially in floodplains. Sometimes it is not difficult to forecast the shifting of a river, and thus spare considerable funds for reinforcement of bridges, roads etc.

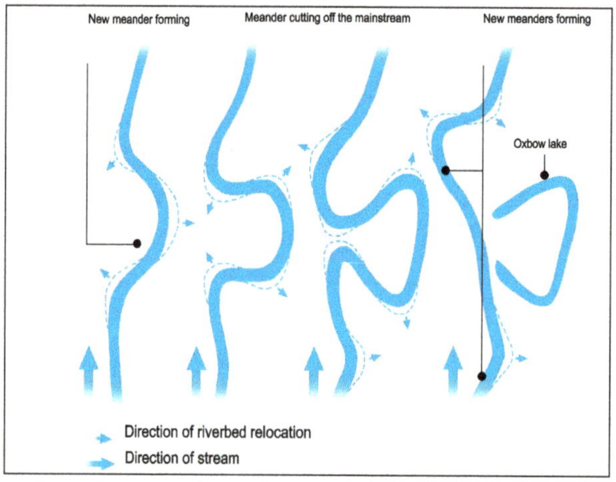

River corridor migration - the same river section is shown in four different successive periods.

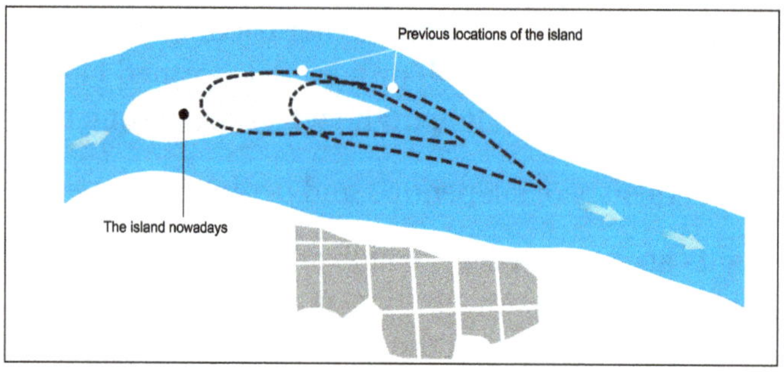

Riparian island relocation.

Riparian Vegetation

Riparian vegetation is a very important component of the river corridor, and its ecosystem significance is even higher. Specific wood species adapted to the river conditions develop along all rivers. With regard to natural rivers, this vegetation forms a continuous belt from the river spring to the mouth. The most typical species in Bulgaria to be found along the streams are willows and alders. River banks in the mountains are covered by green alder. Downstream its place is consecutively replaced by white and black alders. Different species of willow, white and black poplar and elm gradually become present. The sycamore is another riparian tree species typical only for the lower courses of our southern rivers (Struma, Mesta, tributaries of Arda, etc).

Wood canopy along smaller rivers may completely cover and overshadow the riverbed, while this effect is only partial along larger rivers. The overshadowing or "doming" of the water course by riparian vegetation has an extremely strong influence on water temperature and sunlight penetration, which favours the development of phyto-plankton.

Roots of the Black alder.

Riparian vegetation plays an important role in river bank consolidation and prevention of coastal erosion. The roots of the black alder are entangled under the water into specific flat netting which retains even the finest particles of the bank layers. Other riparian species such as willows have similarly entangled roots but less potential to protect against erosion.

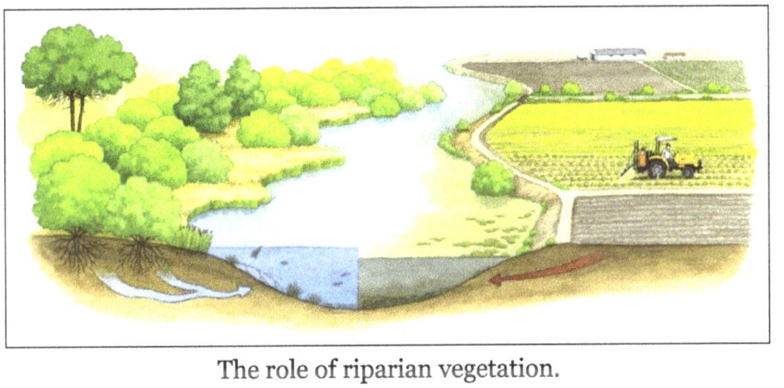

The role of riparian vegetation.

The roots of riparian vegetation also serve as a filter that "swallows" most of the nutrients dissolved in groundwater before reaching the river. Once again, the black alder appears to be a remarkable tree. It is a species that has one of the highest capacities to extract nutrients from groundwater. Therefore sometimes even a thin (10-15 m) riparian belt of alder forest can "clean" some 80% of the nutrient pollutants that would otherwise flow into the river.

Vegetation in catchment basins considerably slows down rainwater runoff to the river, thus decreasing the risk of a high tide. The abundant vegetation controls evaporation from earth surface and keeps groundwater near the surface.

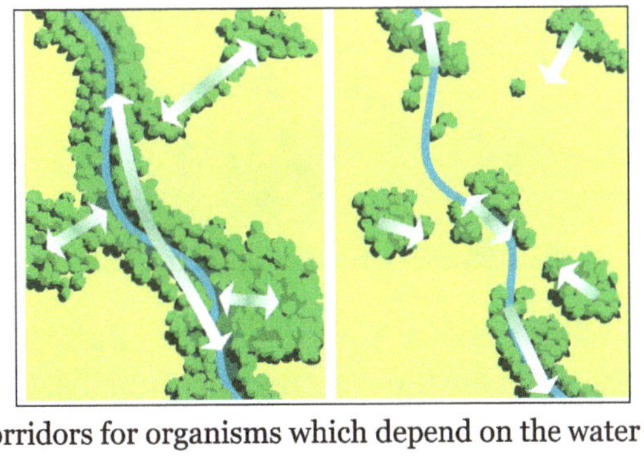

Rivers are migration corridors for organisms which depend on the water and the forest habitats.

River Biocorridors

Rivers are important migration corridors. We know that water wildlife migrates with the purpose of feeding, breeding and to compensate for the river flow both upstream and downstream the river. The possibility to migrate enables water life to avoid unfavourable river sections, such as temporarily polluted places, gravel pits etc. It is even more important for water life to be able to go back and re-colonize an area when conditions are restored. For instance, fish may leave a river section which suffered a severe pollution discharge impact and then restore its population in neighbouring river sections. If there are functional barriers in the biocorridor such as dams, strongly polluted sections, or concrete riverbeds, migration is hindered and the relevant river section may remain with no fish for long time even if the conditions improve.

Another key component of the river biocorridor is the continuous belt of riparian vegetation. This is a basic interlink for many animals in the course of their regular migrations.

In the past, vast forests covered the lowlands. Nowadays forests in the lowlands are fragmented and small. They are dotted amongst vast areas of farmland, urban developments, and highways, which are an overwhelming barrier for forest wildlife.

River corridors are often the only natural link between isolated forest areas and the survival of a large number of plants and animals that depends on them.

Stages of River Development

Youthful River

Perhaps the most dynamic of all rivers is a Youthful River. Rafters seeking an exciting ride will surely gravitate towards a young river for their recreational thrills. Characteristically youthful rivers are found at higher elevations, in mountainous areas, where the slope of the land is steeper.

Water that flows over such a landscape will flow very fast. Youthful rivers can be a tributary of a larger and older river, hundreds of miles away and, in fact, they may be close to the headwaters (the beginning) of that larger river.

- The river flowing down a steep gradient (slope).
- The channel is deeper than it is wide and V-shaped due to downcutting rather than lateral (side-to-side) erosion.
- Its velocity is fast and strong.
- Capable of moving all sediment sizes from ions in solution, to silts and clays, also cobbles and boulders.
- Steep sided cliffs flank the river.
- A floodplain does not exist. There are no grassy areas beside the river where a person can walk.
- Rapids may be present due to the water velocity and the presence of boulders in the channel. Waterfalls are also a feature of a young river.
- Erosion is prominent over deposition.

Mature River

The Mature River is an in-between stage. The river still downcuts though to a much lesser degree than the Youthful River does but it also erodes laterally, though not as extensively, when compared to the Old Age River.

The landscape over which it passes is steep enough that the river's slope enables a velocity capable of moving not only the finer sediments, but also the larger pebbles and cobbles by way of rolling, bouncing and saltation along the river bed. The area through which the river flows may be mountainous but they will not be as high as the Young River's locale. A "hilly" landscape would be a better description for the surrounding area. Rapids are absent and so is the V-shaped channel. The channel of a Mature River is U-shaped but deeper than and not as wide as the Old Age rivers' channel.

Upon observation of a Mature River, here is what one might see:

- The river flows down a moderate gradient (slope).
- The channel is U-shaped and wider than a youthful river yet deeper than an old age channel due to moderate downcutting but also lateral (side-to-side) erosion. In general, the channel is broader with gentler slopes.
- The velocity is greater than an old age river but less than a youthful one.
- Capable of moving many sediment sizes from ions in solution, to silts and clays, also cobbles, but normally not boulders unless peak flooding occurs.
- Cliffs may flank the river at a distance.
- A floodplain exists with grassy areas beside the river along which a person can walk.
- Meanders may be present though they will not be as "curvy" as those found in Old Age Rivers.
- Erosion is present though deposition of sediments also occurs.
- There is more water in the stream channel, i.e. the river has a greater discharge than the youthful river. This means the river is capable of carrying more a greater volume of sediment.

Old Age River

The Mississippi is such a river and its slow movement and suspended sediment make it appear, in some locations, more like a river of mud than a river of water. In fact, the Mississippi River can be so muddy that one may think twice before taking a dip.

That does not mean that an Old Age River cannot have its own dynamic behavior. During flood stages, water that overflows the river's banks, pouring over and beyond levees, attains velocities not only capable of moving large boulders, but also large houses. When Old Age Rivers are filled to capacity due to extensive periods of precipitation and the spring snowmelt around its tributaries, flooding makes the newspaper headlines. Extensive property and agricultural damage, as well as loss of life, are the results of hazardous flooding associated with Old Age Rivers.

Old Age Rivers actually have more distinguishing features to speak of than the Youthful and Mature Rivers do.

Upon observation of an Old Age River, here is what one might see:

- The river flows down a very shallow gradient (slope).
- The channel wider than it is deep with a very broad and U-shape due to extensive lateral (side-to-side) erosion.
- Its velocity is quite slow and that means the river is.
- Capable of moving small-sized sediments, i.e. silts and clays. Small sediments are suspended in the slow-moving water giving the river a "muddy" appearance. Dissolved salts and ions are carried in solution.
- The general landscape surrounding the river is flatter and less steeply sloped. If hilly areas exist, they are further away from the river channel, keep at a distance by the wide floodplain which flanks the river.
- A wide floodplain characterizes Old Age Rivers. There are grassy areas beside the river along which a person can walk. However, the floodplain is often marshy and swampy due to flooding of the river valley. In places, hip boots would be a good idea.
- Curvy "S-shaped" Meanders are abundant and prominent features of an older river.
- Yazoo Streams run parallel to the main river but do not join it.
- Oxbow Lakes exist within the floodplain. Meanders were cut off from the main stream due to extensive erosion and deposition. The meander is now its own lake but, with no water entering to replenish its supply, it will eventually dry up and become a Meander Scar.
- Natural Levees, ridges formed by successive floods that deposit sediment over time, flank the outside meander curves.
- Point bars are areas of deposition on the inside curves of a meander. When water velocity decrease on the inside curves, deposition of sediment occurs, filling in the inside curve over time.
- Fed by many smaller tributaries which join the main river at various locations, the discharge (volume of water), is quite large. This means that, although the river is capable of moving only the finer sediments, it can move large amounts of sediments.
- Erosion is present on the outside curves of meanders while deposition of sediments occurs on the inside curves.

Disruptions to the River Community

Serious disruptions may eliminate large segments of the community. This is what is meant by loss of biodiversity. Human Impacts on River Communities People can affect rivers directly by dumping things in them or changing their channels or indirectly by changing the land through which the rivers flow.

Removing Trees

From the banks means less food from leaf fall, and warmer water temperatures as the sun strikes and heats more of the water surface. Since most biochemical processes speed up as the temperature increases, warmer water can push the system into high gear.

Decomposers work and use oxygen at a faster rate. To compound the problem, warm water holds less oxygen than cold water. Some species can't survive in warm water, either due to lower oxygen levels, or because they are sensitive to heat. Also, loss of large woody debris such as fallen logs and root wads removes hiding and resting areas for fish, known as "cover."

Impervious Road

Parking lot and roof surfaces speed up the over-land flow of water, which means more water gets to the river faster. Higher flood flows during spring and fall, more channel scouring, and more erosion (since there's more energy and soil particle-dislodging potential) are the result. Less water is able to infiltrate the soil to replenish groundwater, so lower summer flows occur. Asphalt surfaces also heat the water entering the river. Pollutants such as oils, trash and even dirt are washed right into the river instead of being filtered through the soil.

Erosion

Now let's add some sediment from eroding crop land, construction sites, logging areas, and bank erosion. Sediment buries the homes and clogs the gills of many aquatic organisms, and catches and holds heat from the sun, causing warmer water temperatures. Often, nutrients and other pollutants, including heavy metals and toxics, attach to sediment, adding to the damage it causes when washed into streams. Habitat alteration from sedimentation is one of the leading causes of fishery decline.

Nutrients

Adding nutrients to a river encourages aquatic plant and algae growth. These plants may create wild swings in oxygen levels, as they release oxygen during the day and along with aquatic animals and decomposers consume it at night. When the plants and algae die and decompose, more oxygen is used. Again, aquatic organisms may suffocate, especially in the dark hours of early morning.

Dam

Changes the physical foundation of the river when it replaces rapids and cascades with a reservoir. Above the dam, water velocity is slowed, causing soil particles and organic material to settle to the bottom and cover the river bed with mud. Oxygen levels are reduced as the organic material, which previously flowed downstream, decomposes. At the same time, oxygen is not replenished as quickly in the reservoir as it was in the rapids.

Downstream of the dam, the flow and oxygen levels may fluctuate dramatically, if water is daily stored and released. The food supply from upstream is reduced. Temperature changes may cause a trout fishery to be succeeded by a warm-water bass population.

Sewage

Means decomposing organic material, which consumes oxygen and adds nutrients to the water. The decomposers go to work. As they work, they breathe in oxygen, breathe out carbon dioxide and create an oxygen demand. If we put too much organic material in, the river can't replace oxygen fast enough and the demand exceeds the supply.

Fish like salmon and trout need high oxygen levels in the stream to survive. So do the egg and juvenile life stages of other fish species. Aquatic organisms that need the most oxygen will suffocate first, reducing the food choices for surviving species. If the oxygen drain continues, more species will perish, reducing the food choices further. In massachusetts, late summer is a stressful time for fish, as water levels drop, water temperatures rise, and dissolved oxygen levels decrease. Added organic or nutrient loads can cause fish kills. Sewage also brings in bacteria and pathogens, making the water unsafe to drink or swim.

Natural Droughts

In late summer can be exacerbated by withdrawals of water from the river directly or through groundwater wells. Shallower waters and lower flows mean less living space is available to fish and other aquatic organisms, increasing competition for food and predation. Lower oxygen levels and increased temperatures stress fish, making them more susceptible to bacterial infections.

Bulldozer

Bulldozer or bucket loader pulling gravel out of the stream changes the physical habitat available for some insect and fish species in addition to causing sedimentation by stirring up the streambed. Not only do these individual actions cause impacts, there is a cumulative impact that may go far beyond the sum total of the individual impacts. Most any river in Massachusetts has been subjected to this treatment at some time in its relationship with humans. This not-so-hypothetical river can no longer support the same aquatic community it once did.

River Restoration

River restoration is the process of managing rivers to reinstate natural processes to restore biodiversity, providing benefits to both people and wildlife. Reintroducing natural processes can reshape rivers to provide the diversity of habitats required for a healthy river ecosystem and ensure their long-term recovery by addressing the root cause of the issue.

Some rivers have been extensively modified to accommodate societal needs for food production, flood protection and economic activity so it is not always possible or desirable to restore to a pristine condition. In this case, improvement of river structure and habitats within the adjacent constraints can increase overall biodiversity and mitigate some issues, as in figure.

"River restoration is the re-establishment of natural physical processes (e.g. variation of flow and sediment movement), features (e.g. sediment sizes and river shape) and physical habitats of a river system (including submerged, bank and floodplain areas)." - International Union for Conservation of Nature 'River Restoration and Biodiversity.

Importance of River Restoration

The fact that most towns and cities developed near rivers illustrates their importance to humans.

Naturally functioning rivers and floodplains provide ample benefits to society including flood regulation, freshwater supply, tourism/recreation, water purification, carbon storage and improved human health. Many of these benefits, along with biodiversity and habitat, are compromised if rivers are modified.

River restoration projects encourage local communities to engage in their local environment, raising awareness of environmental issues. To ensure local communities benefit as much as possible from river restoration projects, it is important to involve all interested organisations and individuals from the outset.

River Restoration Process

Deciding the appropriate river restoration technique to use is dependent on river type, modification extent and adjacent restrictions (e.g. infrastructure).

Managing Catchments

Ideally, to deliver multiple benefits to society and wildlife, river restoration should be considered at catchment scale and prioritised as part of wider catchment plans. The Catchment-Based Approach is a community-led approach that engages individuals and organisations from across society to improve freshwater environments. Catchment management groups are the best place to identify issues and agree restoration priorities. This approach will produce the most beneficial results as the river system can be viewed as a whole - focusing on a specific reach, without a wider catchment understanding, can lead to detrimental effects elsewhere.

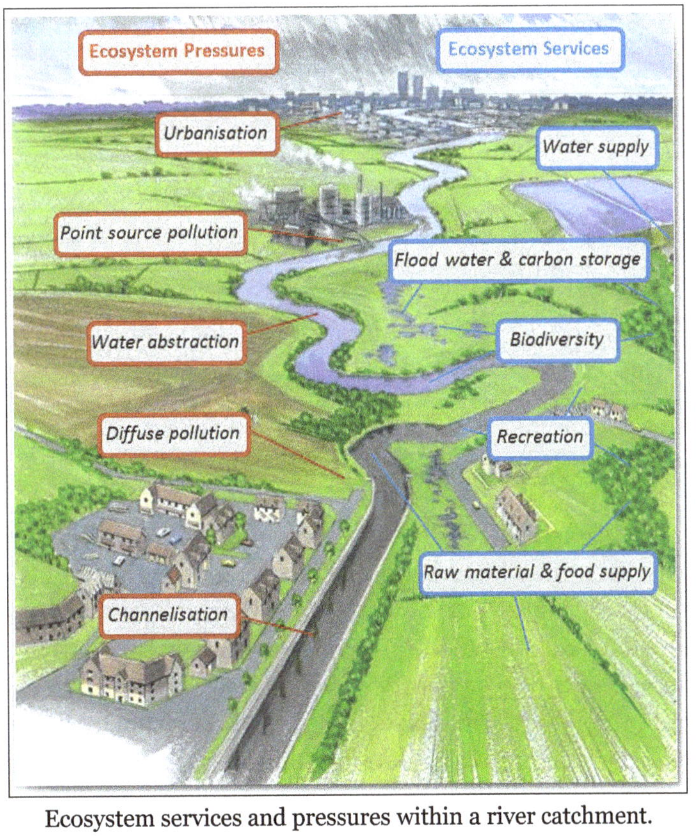

Ecosystem services and pressures within a river catchment.

Restoring a more Natural River Course

River straightening, diverting and over-deepening has been common practice in the past, to create space for land development, enhance river navigation, improve land drainage and reduce flooding. Straightened channels generally lack flow and habitat diversity as their profile has been extensively modified and their features have been removed. Straightening can increase the risk of flooding downstream as water moves faster through the modified section and increases discharge downstream. Remeandering reinstates a more natural course and river profile with the aim of improving habitat diversity and biodiversity. Flow can be returned to the former river course or an entirely new course can be constructed if the old channel cannot be identified or is not accessible.

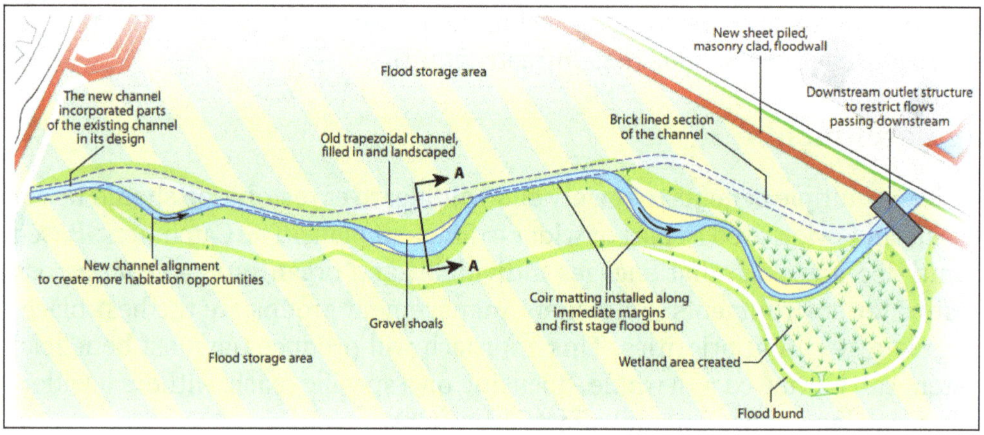

Diagram showing a remeandered channel through Inch Park in Edinburgh - parks offer a good opportunity to remeander as space is less restricted than in residential/urban areas.

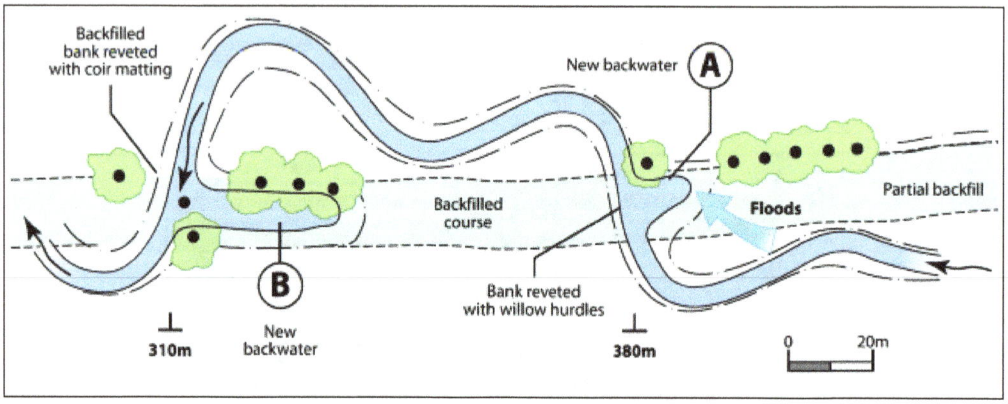

Diagram illustrating the creation of floodplain along the River Thames on Pinkhill Meadow.

Floodplains are an important aspect of a riverine environment, providing flood storage, fish refuge and habitat diversity but often they have been disconnected and drained to protect housing from flooding and create land for agriculture or development. This disconnection and reduction in storage may cause a greater flood risk downstream as water moves through a catchment quicker. To restore connectivity, flood banks can be breached or set back in carefully chosen locations, allowing water to spill out onto the floodplain again. Benefits of reconnecting the floodplain include an increase in flood storage area, recreation of wetland habitat, reintroduction of wetland species and creating refuge for fish during high flows.

In-stream Enhancement

Pictures of berm creation during construction and one year after showing vegetation colonisation on the River Somer.

If floodplains have been reclaimed and developed, which is the case in most urban areas, it is often not feasible to make space for large-scale river restoration. If the river has concrete banks or bed, which is again often the case in urban areas, reinstating some natural processes within the river channel can be the next best option. There are many novel in-stream enhancement techniques to consider but most involve introducing some form of roughness in-channel, such as woody material, reworking gravels or creating berms to create flow diversity, new habitats and areas of refuge. This form of restoration can provide a great opportunity to get local communities involved as it usually requires a hands-on approach.

References

- River-ecosystems, resources: sciencelearn.org.nz, Retrieved 18 April, 2019
- Riverecology-eng: awsassets.panda.org, Retrieved 21 May, 2019
- River-ecology, starting, mwwp: umass , Retrieved 26 July, 2019
- What-is-river-restoration_final, esmee, training, general, default, sites: .therrc.co.uk, Retrieved 7 May, 2019

4

Lake Ecology

The branch of ecology which focuses on the study of still water ecosystems is termed as lake ecology. Some of the aspects studied in relation to this field are watershed ecology and the different lake zones. This chapter closely examines these key concepts of lake ecology as well as the benefits of lakes to provide an extensive understanding of the subject.

Lake ecosystems are influenced by their watersheds, thegeological, chemical and biological processes that occur on the land and streams that lie uphill. The movement of chemicals, sediments, detritus, and of many organisms, is typically unidirectional from the watershed to the lake, butfish may migrate upstream, and aquatic insects may emerge and disperse on to land. A lake and its watershed are often considered to be a single ecosystem. A lake is a body of water that is surrounded by land. There are millions of lakes in the world. They are found on every continent and in every kind of environment—in mountains and deserts, on plains, and near seashores. Lakes vary greatly in size. Some measure only a few square meters and are small enough to fit in your backyard. Such small lakes are often referred to as ponds. Other lakes are so big that they are called seas. The Caspian Sea, in Europe and Asia, is the world's largest lake, with an area of more than 370,000 square kilometers (143,000 square miles). All lakes fill bowl-shaped depressions in the Earth's surface, called basins. Lake basins are formed in several ways. Many lakes, especially those in the Northern Hemisphere, were formed by glaciers that covered large areas of land during the most recent ice age, about 18,000 years ago. The huge masses of ice carved out great pits and scrubbed the land as they moved slowly along. When the glaciers melted, water filled those depressions, forming lakes. Glaciers also carved deep valleys and deposited large quantities of earth, pebbles, and boulders as they melted. These materials sometimes formed dams that trapped water and created more lakes.

Geological Characteristics of Lake

Lake Variability

Lakes are extremely heterogeneous, or patchy. The physical, chemical, and biological characteristics of lakes are extremely variable. Lakes vary physically in terms of light levels, temperature, and water currents. Lakes vary chemically in terms of nutrients, major ions, and contaminants. Lakes vary biologically in terms of structure and function as well as static versus dynamic variables, such as biomass, population numbers, and growth rates. There is a great deal of spatial heterogeneity in all these variables, as well as temporal variability on the scales of minutes, hours, diel (day/night), seasons, decades, and geological time. Though lakes vary in many dimensions they are actually highly structured, similar to a forest ecosystem where, for example, a variety of physical variables (light, temperature, moisture) vary from the soil up through the canopy.

Light

Perhaps the most fundamental set of properties of lakes relates to the interactions of light, temperature and wind mixing. The absorption and attenuation of light by the water column are major factors controlling temperature and potential photosynthesis. Photosynthesis provides the food that supports much of the food web. It also provides much of the dissolved oxygen in the water. Solar radiation is the major source of heat to the water column and is a major factor determining wind patterns in the lake basin and water movements.

Light intensity at the lake surface varies seasonally and with cloud cover and decreases with depth down the water column. The deeper into the water column that light can penetrate, the deeper photosynthesis can occur. Photosynthetic organisms include algae suspended in the water (phytoplankton), algae attached to surfaces (periphyton), and vascular aquatic plants (macrophytes).

The rate at which light decreases with depth depends upon the amount of light-absorbing dissolved substances (mostly organic carbon compounds washed in from decomposing vegetation in the watershed) and the amount of absorption and scattering caused by suspended materials (soil particles from the watershed, algae and detritus). The percentage of the surface light absorbed or scattered in a 1 meter long vertical column of water, is called the vertical extinction coefficient. This parameter is symbolized by "k".

In lakes with low k-values, light penetrates deeper than in those with high k-values. Figure shows the light attenuation profiles from two lakes with attenuation coefficients of 0.2/m and 0.9/m.

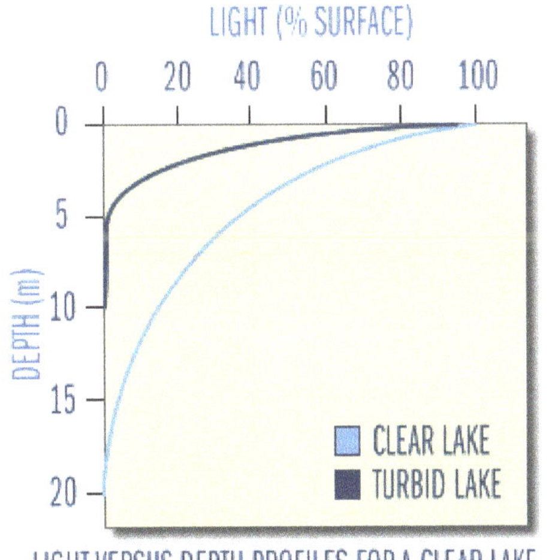

LIGHT VERSUS DEPTH PROFILES FOR A CLEAR LAKE ($k=0.2\ M^{-1}$) AND A TURBID LAKE ($k=0.9\ M^{-1}$).

The maximum depth at which algae and macrophytes can grow is determined by light levels. Limnologists estimate this depth to be the point at which the amount of light available is reduced to 0.5%–1% of the amount of light available at the lake surface. This is called the euphotic zone. A general rule of thumb is that this depth is about 2 to 3 times the limit of visibility as estimated using a Secchi disk. Light may be measured in a variety of ways for a number of different characteristics. The reader is referred to the reference texts for more information. Since photosynthesis depends fundamentally on light, significant changes in light penetration in a lake will produce a variety of direct and indirect biological and chemical effects. Significant changes in lake transparency are most often the result of human activities, usually in association with land use activities in the watershed.

Table: Estimated ranges of water transparency values for various lakes.

LAKE	k (m^{-1})	Secchi Depth (m)	Euphotic Zone (m)	Description
Crater Lake (OR)	0.06—0.12	25—45	>120	Clear, sky blue ultra-oligotrophic lake
Lake Tahoe (CA/NV)	0.12	40	90—136	As above but decreasing clarity since1960s due to watershed overdevelopment
		Lake Superior		
Lake Superior (Blue water)	0.13	15—20	46—60	Ultra-oligotrophic; most oligotrophic of the Laurentian Great Lakes
Lake Superior (Green water near Duluth)	0.3	5—12	20—30	Western arm near Duluth and St. Louis River and harbor inputs
St. Louis River (Duluth-Superior Harbor)	4.21	0.7	>5	Brown (bog) stained from river plus high suspended sediments
Lake Michigan	0.19—0.24	?	19—31	Meso-oligotrophic
Lake Huron	0.1 — 0.5	?	25—31	Meso-oligotrophic
Lake Erie	0.2 — 1.2	2—10 (1970—1990) >10 (1993—1995)	12—26	Eutrophic (clarity improving recently due to zebra mussels)
Lake Ontario	0.15 — 1.2	?	12—29	Mesotrophic
Lake Baikal, Siberia	0.2	5—40	15—75	Oligotrophic
Grindstone Lake	0.82	3—6	8—20	Mesotrophic, water is fairly stained or colored
(Pine County, MN) Ice Lake	0.83	2—5	6—15	Mesotrophic
	(Itasca County, MN) Lake Minnetonka (Hennepin County, MN)			
West Upper	0.78	1.4	3—5	Mesotrophic
Halsted Bay	2.9	0.5	<2	Eutrophic

"Clear" water refers to the lack of bog-staining color. Vertical extinction coefficient, k, defined previously. Secchi depths in meters.

Density Stratification

In the spring, immediately after ice-out in temperate climates, the water column is cold and nearly isothermal with depth. The intense sunlight of spring is absorbed in the water column, which also heats up as the average daily temperature of the air increases. In the absence of wind, a temperature profile with depth might be expected to resemble figure, decreasing exponentially with depth. However, density, another physical characteristic of water, plays an important role in modifying this pattern.

Water differs from most other compounds because it is less dense as a solid than as a liquid. Consequently ice floats, while water at temperatures just above freezing sinks. As most compounds change from a liquid to a solid, the molecules become more tightly packed and consequently the compound is denser as a solid than as a liquid. Water, in contrast, is most dense at 4 °C and becomes less dense at both higher and lower temperatures. The density/temperature relationship of fresh water is shown in figure. Because of this density-temperature relationship, many lakes in temperate climates tend to stratify, that is, they separate into distinct layers.

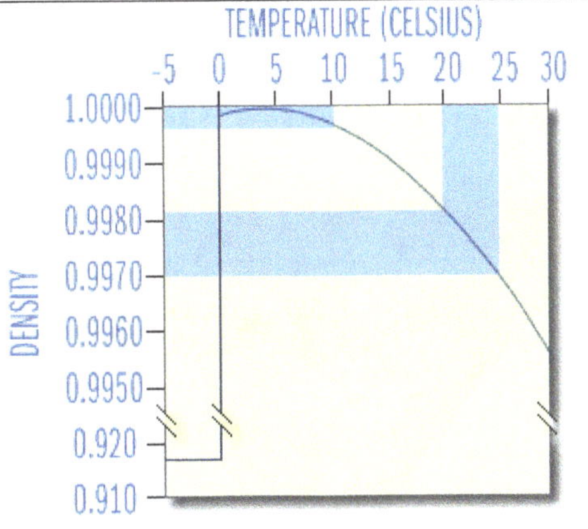

DENSITY/TEMPERATURE RELATIONSHIP FOR DISTILLED WATER. SHADED AREAS SHOW RELATIVE DIFFERENCE IN DENSITY FOR 5°C TEMPERATURE CHANGES.

Spring

In lakes of the upper Midwest and at higher elevations, the water near a lake's bottom will usually be at 4 °C just before the lake's ice cover melts in the spring. Water above that layer will be cooler, approaching 0 °C just under the ice. As the weather warms, the ice melts. The surface water heats up and therefore it decreases in density. When the temperature (density) of the surface water equals the bottom water, very little wind energy is needed to mix the lake completely. This is called turnover. After this spring turnover, the surface water continues to absorb heat and warms.

As the temperature rises, the water becomes lighter than the water below. For a while winds may still mix the lake from bottom to top, but eventually the upper water becomes too warm and too buoyant to mix completely with the denser deeper water. As suggests, the relatively large differences in density at higher temperatures are very effective at preventing mixing. It simply takes too much energy to mix the water any deeper.

Summer

As summer progresses, the temperature (and density) differences between upper and lower water layers become more distinct. Deep lakes generally become physically stratified into three identifiable layers, known as the epilimnion, metalimnion, and hypolimnion. The epilimnion is the upper, warm layer, and is typically well mixed. Below the epilimnion is the metalimnion or thermocline region, a layer of water in which the temperature declines rapidly with depth. The hypolimnion is the bottom layer of colder water, isolated from the epilimnion by the metalimnion. The density change at the metalimnion acts as a physical barrier that prevents mixing of the upper and lower layers for several months during the summer.

The depth of mixing depends in part on the exposure of the lake to wind (its fetch), but is most closely related to the lake's size. Smaller to moderately-sized lakes (50 to 1000 acres) reasonably may be expected to stratify and be well mixed to a depth of 3–7 meters in north temperate climates. Larger lakes may be well mixed to a depth of 10–15 meters in summer (e.g., Western Lake Superior near Duluth, MN).

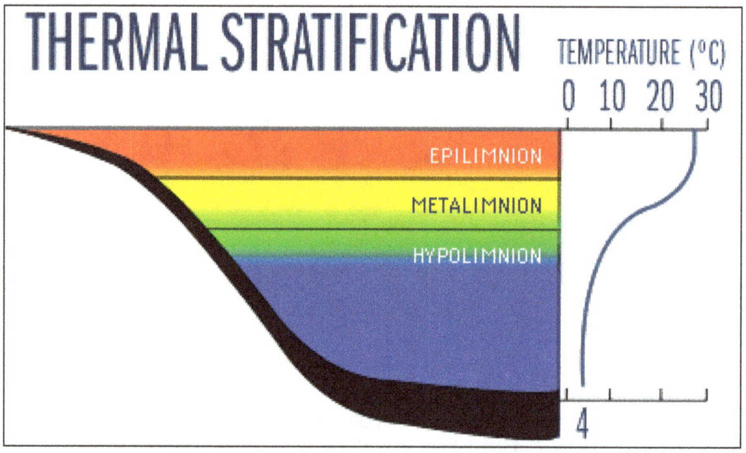

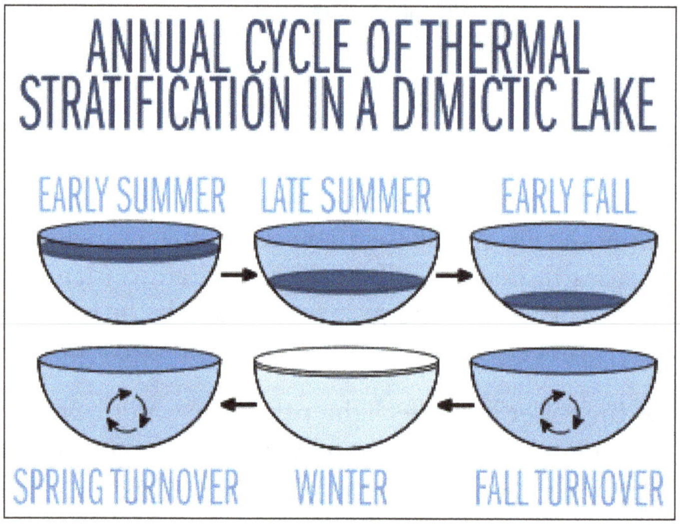

Autumn

As the weather cools during autumn, the epilimnion cools too, reducing the density difference between it and the hypolimnion. As time passes, winds mix the lake to greater depths, and the thermocline gradually deepens. When surface and bottom waters approach the same temperature and density, autumn winds can mix the entire lake; the lake is said to "turn over." As the atmosphere cools, the surface water continues to cool until it freezes. A less distinct density stratification than that seen in summer develops under the ice during winter. Most of the water column is isothermal at a temperature of 4 °C, which is denser than the colder, lighter water just below the ice. In this case the stratification is much less stable, because the density difference between 0 °C and 4 °C water is quite small. However, the water column is isolated from wind-induced turbulence by its cap of ice.

This pattern (spring turnover — summer stratification — fall turnover — winter stratification) is typical for temperate lakes. Lakes with this pattern of two mixing periods are referred to as dimictic. Many shallow lakes, however, do not stratify in the summer, or stratify for short periods only, throughout the summer. Lakes that stratify and destratify numerous times within a summer are known as polymictic lakes.

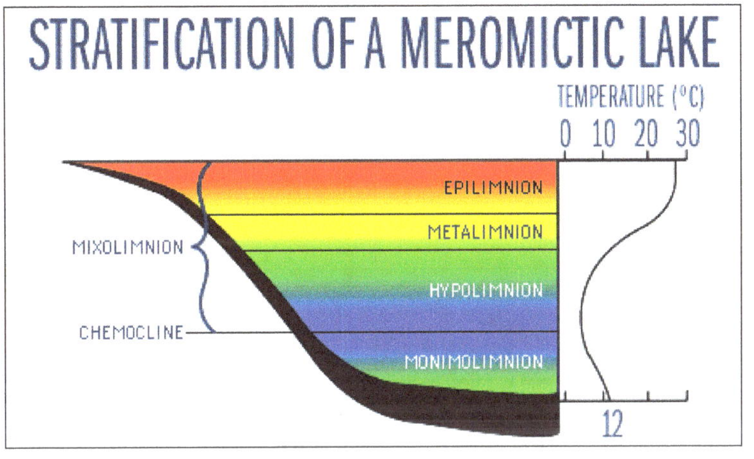

The West Upper Lake Station of Lake Minnetonka, Lake Independence and thousands of other Upper Midwestern lakes that are relatively deep (>10 meters) and reasonably large (>100-200 acres or 40-80 ha) are probably dimictic, leading to complete re-oxygenation of the water column for at least some period of time. Ice Lake, though small (41 acres or 16.6ha), is sheltered and deep (16 m) for its size. Lakes that have formed in former open pit mines in Northeastern Minnesota are unusually deep for their size. Lakes with these characteristics probably only mix completely once a year in the fall for a brief period before freezing. Some of the deeper mine pit lakes (>75 meters deep) probably never mix completely to the bottom, although data are sparse.

Much less common are lakes that circulate incompletely resulting in a layer of bottom water that remains stagnant. To distinguish them from the holomictic (mixing from top to bottom) lakes, these partially mixing lakes are referred to as meromictic. They mix partially, in the sense that they may have extensive mixing periods which go quite deeply into the hypolimnion, but they do not turn over completely, and a layer of bottom water remains stagnant and anoxic for years at a time. The non-mixing bottom layer is known as the monimolimnion and is separated from the mixolimnion (the zone that mixes completely at least once a year) by the chemocline. The stagnant, and typicallyanaerobic, monimolimnion has a high concentration of dissolved solids compared to the mixolimnion. In general, meromictic lakes have large relative depths. These lakes are typically small and sheltered from the wind by the morphology of their basin. In this case, the density differences caused by temperature are smaller than density differences due to the high dissolved solids (salts) concentration of the monimolimnion. Large lakes that rarely freeze over are also typically monomictic, mixing throughout the fall, winter and spring and stratifying in the summer.

Mixing Regime	Lakes	Max Depth (M)	Area (Acres/Hectares)
Dimictic (2mixes/yr)	Lake Minnetonka (Minneapolis, MN)	34	14,004 acres (5,670 ha)
	Grindstone Lake (Sandstone, MN)	46	500 acres (200 ha)
	Lake Independence (Minneapolis, MN)	18	850 acres (344 ha)
	Pike Lake (Duluth, MN)	18	8 500 acres (200 ha)
Monomictic (1mix/yr – mixed all winter and spring)	Lake Erie	70	6.4 x 106 acres (2.6 x 106 ha)
	Lake Huron	228	14.8 x 106 acres (6.0 x 106 ha)
	Ice Lake* (Grand Rapids, MN)	16	41 acres (16.6 ha)
	Lake Michigan	281	14.4 x 106 acres (5.8 x 106 ha)
	Lake Ontario	244	4.9 x 106 acres (2.0 x 106 ha)
	Lake Superior	300	20.3 x 10^6 acres (8.2 x10^6 ha)
	Lake Tahoe (CA/NV)	499	123,253 acres (49,900 ha)
	Lake Mead (NV – largest US reservoir)	180	163,320 acres (66,096 ha)
Polymictic (many mixes/yr)	Shallow lakes & ponds	< 4	wide range
	Mille Lacs Lake, MN	13	132,510 acres (53,648 ha)
	St. Louis River and Duluth-Superior Harbor	1-8	11,993 acres (4,856 ha)
Meromictic (never totally mixed because of stagnant bottom layer)	Miners Pit Lake (Ely, MN)	48	138 acres (56 ha)
	Pennington Pit Lake, (Crosby, MN)	79	57 acres (23 ha)
	Brownie Lake (Minneapolis, MN)	15	18 acres (7.3 ha)
	Deming Lake (Itasca State Park, MN)	14	12.3 acres (5.0 ha)
	Big Soda Lake (Fallon, NV)	60	400 acres (160 ha)

General Lake Chemistry

In the absence of any living organisms, a lake contains a wide array of molecules and ions from the weathering of soils in the watershed, the atmosphere, and the lake bottom. Therefore, the chemical composition of a lake is fundamentally a function of its climate (which affects its hydrology) and its basin geology. Each lake has an ion balance of the three major anions and four major cations.

Table: Ion Balance for Typical Fresh Water.

Anions	Percent	Cations	Percent
HCO_3^-	73%	Ca^{+2}	63%
SO_4^{-2}	16%	Mg^{+2}	17%
Cl^-	10%	Na^+	15%
		K^+	4%
other	<1%	other	< 1%

Ion balance means the sum of the negative ions equals the sum of the positive cations when expressed as equivalents. These ions are usually present at concentrations expressed as mg/L ppm whereas other ions such as the nutrients phosphate, nitrate, and ammonium are present at μg/L ppb levels.

Humans can have profound influences on lake chemistry. Excessive landscape disturbance causes higher rates of leaching and erosion by removing vegetative cover, exposing soil, and increasing water runoff velocity. Lawn fertilizers, wastewater and urban storm water inputs all add micronutrients such as nitrogen and phosphorus, major ions such as chloride and potassium, and, in the case of highway and parking lot runoff, oils and heavy metals. Emissions from motorized vehicles, fossil fuel-burning electric utilities and industry, and other sources produce a variety of compounds that affect lake chemistry.

Perhaps the best understood ions are H^+ (hydrogen ion, which indicates acidity), $SO4^{-2}$ (sulfate) and NO^- (nitrate) which are associated with acid rains. Mercury (Hg) is another significant air pollutant affecting aquatic ecosystems and can bio accumulate in aquatic food webs, contaminating fish and causing a threat to human and wildlife health.

Lakes with high concentrations of the ions calcium (Ca^{+2}) and magnesium (Mg^{+2}) are called hardwater lakes, while those with low concentrations of these ions are called softwater lakes. Concentrations of other ions, especially bicarbonate, are highly correlated with the concentrations of the hardness ions, especially Ca^{+2}. The ionic concentrations influence the lake's ability to assimilate pollutants and maintain nutrients in solution. For example, calcium carbonate ($CaCO_3$) in the form known as marl can precipitate phosphate from the water and thereby remove this important nutrient from the water.

The total amount of ions in the water is called the TDS (total dissolved salt, or total dissolved solids concentration). Both the concentration of TDS and the relative amounts or ratios of different ions influence the species of organisms that can best survive in the lake, in addition to affecting many important chemical reactions that occur in the water.

One example of particular interest in the Great Lakes region involves the calcium requirement of the exotic zebra mussel that is causing profound changes in Lake Erie. Lake Superior appears to be relatively immune to infestation by this invader because of low calcium concentration. Its bays, however, such as the lower St. Louis River and Duluth-Superior Harbor, may not be immune to zebra mussel infestation.

Dissolved Oxygen

Biological activity peaks during the spring and summer when photosynthetic activity is driven by high solar radiation. Furthermore, during the summer most lakes in temperate climates are stratified. The combination of thermal stratification and biological activity causes characteristic patterns in water chemistry. Shows the typical seasonal changes in dissolved oxygen (DO) and temperature. The top scale in each graph is oxygen levels in mg O_2/L. The bottom scale is temperature in °C. In the spring and fall, both oligotrophic and eutrophic lakes tend to have uniform, well-mixed conditions throughout the water column. During summer stratification, the conditions in each layer diverge.

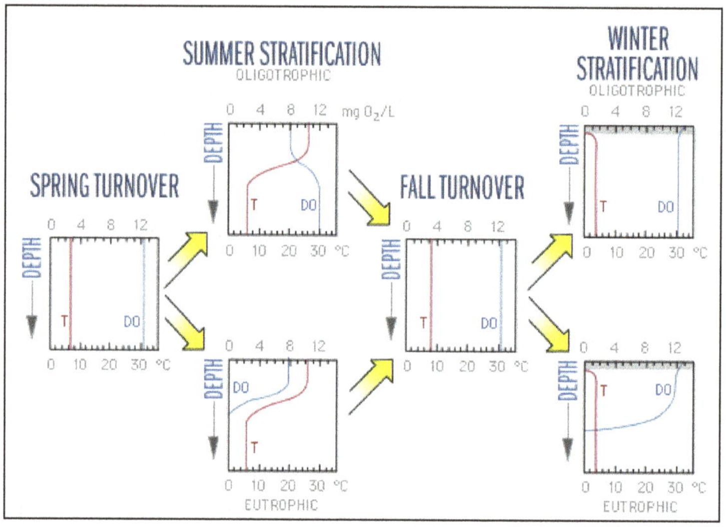

The DO concentration in the epilimnion remains high throughout the summer because of photosynthesis and diffusion from the atmosphere. However, conditions in the hypolimnion vary with trophic status. In eutrophic (more productive) lakes, hypolimnetic DO declines during the summer because it is cut-off from all sources of oxygen, while organisms continue to respire and consume oxygen. The bottom layer of the lake and even the entire hypolimnion may eventually become anoxic, that is, totally devoid of oxygen. In oligotrophic lakes, low algal biomass allows deeper light penetration and less decomposition. Algae are able to grow relatively deeper in the water column and less oxygen is consumed by decomposition. The DO concentrations may therefore increase with depth below the thermocline where colder water is "carrying" higher DO leftover from spring mixing (recall that oxygen is more soluble in colder water). In extremely deep, unproductive lakes such as Crater Lake, OR, Lake Tahoe, CA/NV, and Lake Superior, DO may persist at high concentrations, near 100% saturation, throughout the water column all year. These differences between eutrophic and oligotrophic lakes tend to disappear with fall turnover.

In the winter, oligotrophic lakes generally have uniform conditions. Ice-covered eutrophic lakes, however, may develop a winter stratification of dissolved oxygen. If there is little or no snow cover to block sunlight, phytoplankton and some macrophytes may continue to photosynthesize, resulting in a small increase in DO just below the ice. But as microorganisms continue to decompose material in the lower water column and in the sediments, they consume oxygen, and the DO is depleted. No oxygen input from the air occurs because of the ice cover, and, if snow covers the ice, it becomes too dark for photosynthesis. This condition can cause high fish mortality during the winter, known as "winter kill." Low DO in the water overlying the sediments can exacerbate water quality deterioration, because when the DO level drops below 1 mg O2/L chemical processes at the sediment-water interface frequently cause release of phosphorus from the sediments into the water. When a lake mixes in the spring, this new phosphorus and ammonium that has built up in the bottom water fuels increased algal growth.

Lake Organisms

Creatures that Go Where they Choose		
Fish	Amphibians	Larger Zooplankton (E.G., Mysids)
Creatures that Go where the Water takes them		
Living Creatures = Plankton Animals - Zooplankton Algae - Phytoplankton Bacteria - Bacterioplankton	Dead Creatures = Detritus Internal - Produced Within Lake External - Washed In From Watershed	
Creatures that Live on the Lake Bottom		
Benthos = Animals Aquatic Insects Molluscs – Clams, Snails Other Invertebrates – Worms, Crayfish	Plants Higher Plants – Macro-phytes Attached Algae – Periph-yton	Bacteria and Fungi Sewage Sludge Aufwuchs - Mixture Of Algae, Fungi And Bacteria

Food Web

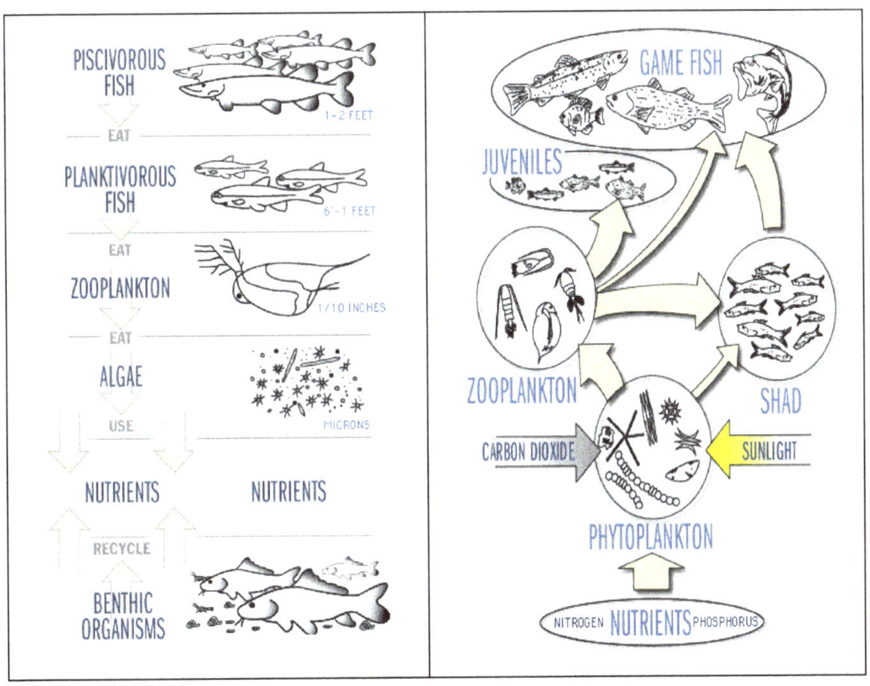

The biological communities within lakes may be organized conceptually into food chains and food webs to help us understand how the ecosystem functions. The simplest illustration of the organization of the organisms within an ecosystem is the ecological pyramid. The broad base of primary producers supports overlying levels of herbivores (zooplankton), planktivores and much smaller numbers of carnivores (predators).

These individual trophic levels may be idealized as a food chain, but in fact many organisms are omnivorous and not necessarily characterized by a particular level. Further, consumers in particular often shift levels throughout their life cycle. For example, a larval fish may initially eat fine particulate material that includes algae, bacteria and detritus.

Then it may switch and graze on larger zooplankton and ultimately end up feeding on so called "forage fish" or even young game fish (i.e., top predators) when it reaches maturity.

Food webs may be described in terms of both energy and nutrient (carbon, nitrogen or phosphorus) flows. Although the process typically begins with sunlight-driven photosynthesis by algae and plants, balanced nutrition is also required to sustain life.

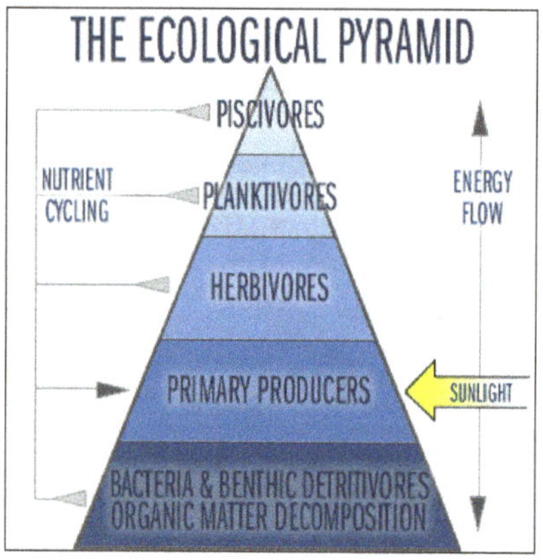

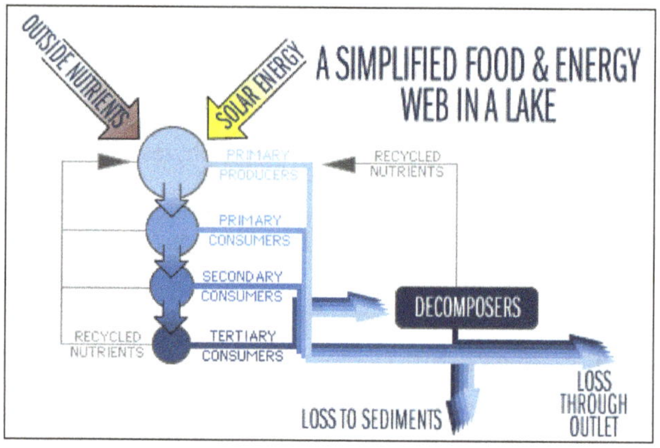

There are two basic life-sustaining processes in lakes, just as on land; photosynthesis and respiration. Green plants capture energy from sunlight to convert nonliving, inorganic chemicals (carbon dioxide, water, and mineral compounds) into living, organic plant tissue. Lake photosynthesizers include algae and macrophytes. Together, they are the primary producers, because they create the organic material required by most other organisms for nutrients and energy. Oxygen, the waste product of photosynthesis, adds to the oxygen supplied to the lake by the atmosphere. In water layers where photosynthetic rates are very high, such as during an algal bloom, the water may become supersaturated. That is, the oxygen content may exceed 100% of saturation with respect to the amount the water could hold if it was allowed to equilibrate with the atmosphere. This saturation value, in turn, depends on the temperature of the water. Colder water can hold more O_2 than warmer water. During periods of stratification, the only potential source of O_2 to the deeper zones of the lake is photosynthesis. This occurs only if light penetrates below the thermocline. In lakes where light does not penetrate below the thermocline, there is no internal source of oxygen to the deeper waters.

Besides light, algae and higher plants need oxygen, carbon dioxide (CO_2), and mineral nutrients to survive and grow. Except for a very few species of blue green algae, most are unable to survive in anoxic (no O_2) water. CO_2 is virtually always available and comes from the weathering of carbonate rocks, such as limestone, in the watershed, diffusion from the atmosphere (very important in soft-water, acid rain sensitive lakes), and from the respiration of organic matter by all of the organisms in the lake. Dissolved mineral nutrients are absorbed from the water by algae and from the water and the sediments by higher plants. Typically, the most important nutrients are phosphorus and nitrogen, because they are present in very low concentrations unless there are sources of pollution and are typically low enough to limit the growth of algae. Other minerals essential to life, such as the major ions (calcium, magnesium, sodium, and potassium) and certain trace metals (iron, cobalt, molybdenum, manganese, copper, boron, and zinc), are usually present at sufficient concentrations. Silicon is required by diatoms and a few other groups of algae and is usually, though not always, present at sufficient levels. Another mineral required by all living things, sulfur (in the form of sulfate), is typically not deficient in lakes.

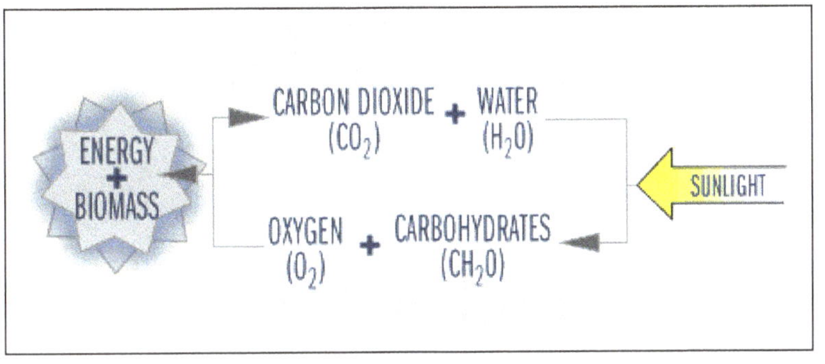

The whole interaction of photosynthesis and respiration by plants, animals, and microorganisms represents the food web. Food webs are usually very complex and, in any one lake ecosystem, hundreds of different species can be involved. Because the available energy decreases at each trophic level, a large food base of primary producers (mostly plants) is necessary to support relatively few large fish.

These plants may die and decompose or be eaten by primary consumers – the second trophic level. This link in the food chain typically involves zooplankton grazing on algae but also includes larval fish eating zooplankton and a variety of invertebrates that eat attached algae (periphyton) and higher plants. Other animals, such as small fish, secondary consumers (third trophic level) eat the primary consumers and thus are considered secondary consumers. Still larger consumers such as large fish, ospreys, and people are tertiary consumers (fourth trophic level). Thus, energy and nutrients originating from the photosynthetic production of biomass and energy cascade through the food web. There is some recycling of nutrients back up to the top of the cascade. Respiration, the oxidation of organic material, releases the energy that was originally captured from sunlight by photosynthesis. Both plants and animals respire to sustain their lives, and in doing so, consume oxygen. Microorganisms (bacteria and fungi) consume a large fraction of available oxygen in the decomposition of excreted and dead organic material.

Decomposers are sinks for plant and animal wastes, but they also recycle nutrients for photosynthesis. The amount of dead material in a lake far exceeds the living material. Detritus is the organic fraction of the dead material, and can be in the form of small fragments of plants and animals or as dissolved organic material. In recent years, scientists have recognized that zooplankton grazing on detritus and its associated bacterial community represent an additional important trophic pathway in lakes.

Primary Producers

Examples of EFM species known to produce anticancer and antioxidative natural products. a Centella asiatica, b Nelumbo nucifera, c Nasturtium officinale, d Ipomoea aquatica, e Ludwigia ad

Much of modern limnological study revolves around the primary productivity of lakes. The ecology of plant growth is of great importance to the character and history of lakes and to all other organisms that live in lakes. The major threat to lakes involves the excessive growth of primary producers due to nutrient inputs caused by poor landuse management. Therefore, it is worth a closer look at these organisms.

Lythrum salicaria (Purple Loosestrife)

The littoral zone is defined by the growth of rooted and floating aquatic plants, or macrophytes. Provides examples of common macrophytes found in Minnesota lakes. The macrophyte community can also include large algae, such as Chara, Nitelle, or Cladophora. In shallow, clear lakes, macrophytes may represent most of the green plant material present and may account for most of the photosynthesis.

There may be few macrophytes in a lake when the bottom is too rocky or too sandy for the plants to anchor themselves, wave action too severe, or the water too deep. Also, sunlight may not reach the bottom even in shallow areas if the concentration of algae or silt is high.

Algae constitute the other main group of primary producers. They come in countless forms and live in nearly all kinds of environments. Most are microscopic, growing as single cells, small colonies, or filaments of cells. Suspended algae are called phytoplankton, while attached algae are called periphyton. Phytoplankton grow suspended in open water by taking up nutrients from the water and energy from sunlight. If their populations are dense, the water will become noticeably green or brown and will have low transparency.

Phytoplankton are classified into groups by the type of pigments they use to perform photosynthesis. While chlorophyll-a is common to all groups there are many other accessory pigments that allow the algae to capture different types of light. Green algae are considered the most closely related to higher plants. Within this group alone there is a great diversity of size, shape, and growth form (single celled, colonial, filamentous, and flagellated). Diatoms belong to a large group, classified as the golden-brown algae, which also includes chrysophytes and dinoflagellates. The most striking characteristic of diatoms and chrysophytes is the ability to form silica (glass) cell walls. Diatoms cell walls are similar to a petri dish, having two halves that fit together. Some chrysophytes have elaborate silica scales, spines, or vase-like shells called loricas. Diatoms are non-motile (unable to swim), so they depend on water turbulence to remain suspended. Chrysophytes have flagella (whip-like appendages) that allow them to control their position in the water column. There are other important algal groups containing motile forms.

Dinoflagellates are another group of golden-brown algae that also have flagella. These cells are capable of moving very rapidly; positioning themselves where light and nutrients are optimal for growth. Another flagellated group called the cryptomonads are very small algae and contain pigments that enable them to photosynthesize under very low light conditions, either very deep in the water column or during those times of the year when sunlight isn't very strong.

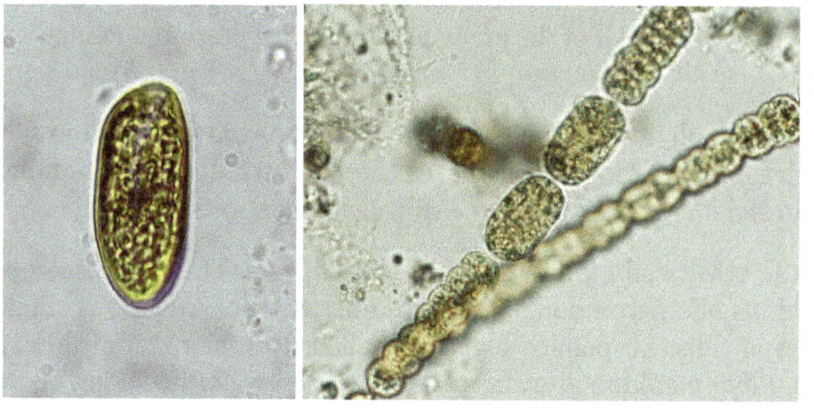

Blue-green "algae" are technically referred to as cyanobacteria since, except for their chlorophyll-based photosynthesis, they are bacteria. They generally receive the greatest amount of research and management attention because of their ability to form nuisance blooms in eutrophic lakes. It is important to remember, however, that blue-green algae are very important primary producers in both freshwater and marine systems, despite often being a nuisance.

Blue-greens have several characteristics that often enable them to dominate and create nuisance or noxious conditions. Some blue-green species have the ability to adjust their buoyancy. They can float or sink depending on light conditions and nutrient supply. All plants, including all algae, typically satisfy their nitrogen requirement by absorbing nitrate (NO_3^-) and/or ammonium (NH_4^+) from the water. However, some blue-greens can fix molecular nitrogen (N_2) derived from the atmosphere and dissolved in the water and convert it to ammonium in the cell through a process called nitrogen fixation. This allows them to maintain high rates of growth when other forms of nitrogen are sufficiently depleted to limit growth by other types of algae. Blue-green algae typically are well-adapted to phosphorus deficiency because of their ability to absorb and store excess phosphorus when it is available - enough to last days to weeks in some cases.

Unlike the green algae and diatoms, the blue-green algae are less suitable food for primary consumers. This is partly because some blue-greens can form large colonies of cells embedded in a gelatinous matrix which may pose handling problems for grazers. They also may produce chemicals that inhibit grazers or makes them "taste bad" to the grazers. Consequently, blue-greens have advantages over other algae at using nutrient and light resources, as well as avoiding being eaten.

Aphanizomenon flos-aquae is a common species of filamentous blue-green algae with the ability to regulate its buoyancy, fix nitrogen, form large inedible colonies, and form algal blooms. Other common bloom genera are Anabaena (N_2-fixing filamentous algae) and Microcystis (colonial; not a N_2-fixer). These bloom-forming algae are known to produce toxins in farm ponds that can poison cattle and, more recently, have been found to produce potent neurotoxins and hepatotoxins that may be a greater public health concern than previously realized.

Chlorophyll – A Measure of Algae

An in-depth microscopic enumeration of the dozens of species of algae present in a water column each time a lake is sampled is prohibitively costly and technically impossible for most monitoring programs. Further, in many lakes a large portion of the algal biomass may be unidentifiable by most experts (these are appropriately called LRGTs or LRBGTs - little round green things and little round blue-green things). However, measuring the concentration of chlorophyll-a is much easier and provides a reasonable estimate of algal biomass. Chlorophyll-a is the green pigment that is responsible for a plant's ability to convert sunlight into the chemical energy needed to fix CO_2 into carbohydrates. To measure chlorophyll-a, a volume of water from a particular depth is filtered through a fine glass-fiber filter to collect all of the particulate material greater than about 1 micron (1/1000th of a millimeter) in size. The chlorophyll-a in this material is then extracted with a solvent (acetone or alcohol) and quantified using a spectrophotometer or a fluorometer.

Both chlorophyll-a and secchi depth are long-accepted methods for estimating the amount of algae in lakes. Secchi depth is much easier and less expensive to determine. However, care must be used in interpreting secchi data because of the potential influence of non-algal particulate material, such as silt from stream discharge or re-suspended bottom sediment. Also, the tea color of some lakes that's due to dissolved organic matter from bogs, can have an effect on secchi depth readings as well. Even if chlorophyll-a is measured, it may be important to also examine the algal community microscopically on occasion, since the mix of species may influence lake management decisions.

Algal Succession

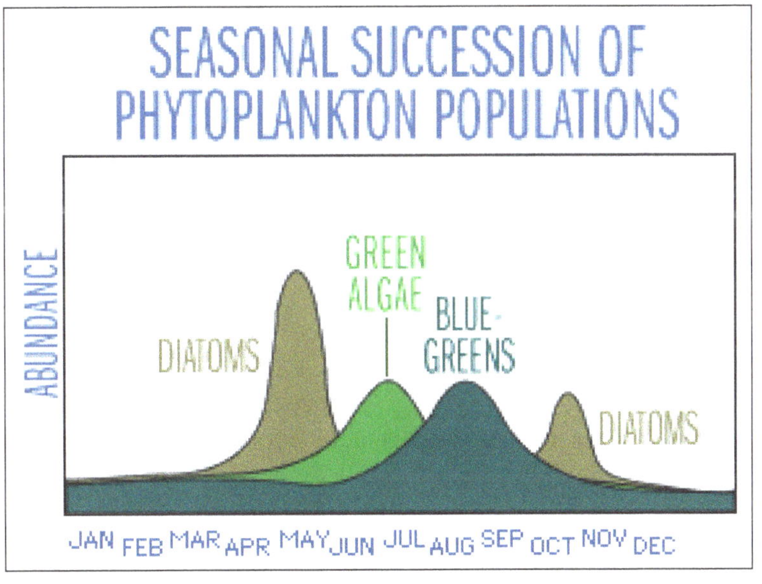

A lake's biological characteristics are determined in large part by physical characteristics of the water column. Important physical characteristics include temperature, light transparency, and wave action, as well as the total abundance of inorganic nutrients, which is largely a watershed characteristic. In addition, preceding populations influence successive populations by assimilating critical nutrients. Populations also have varying susceptibilities to grazing by zooplankton, which vary seasonally in type and abundance. As physical, chemical, and biological conditions in the lake change over time, some species will be effectively eliminated from a lake because they cannot tolerate the new conditions. Other species will be out-competed by organisms that are better adapted to the new environment.

These changes represent an important ecological pattern in lakes known as algal succession. In most natural systems the seasonal succession of algae (and macrophytes) is a recurrent, if not exactly repetitive, yearly cycle. A typical algal succession is shown in figure. Some species flourish for a period of time and then give way to other species more compatible with changed conditions, such as warmer water, more daylight, or lower concentrations of phosphorus or nitrogen. Short-lived plankton communities are characterized by these seasonal fluctuations; longer-lived organisms, such as fish, must be tolerant of lake conditions all year.

Consumers

Zooplankton, small animals that swim about in open water, are primary consumers. They graze on algae, bacteria, and detritus (partially decayed organic material). Some species can be seen with the naked eye, although they are more easily observed with a hand lens or low-power microscopes. If you wish to see them, stare into the water of a pond or lake on a calm night with a flashlight beam shining from above.

Secondary consumers, such as planktivorous fish or predaceous invertebrates, eat zooplankton. While photosynthesis limits plant growth to the sunlit portions of lakes, consumers can live and grow in all lake zones, although the lack of oxygen (anoxia) may limit their abundance in bottom waters and sediments.

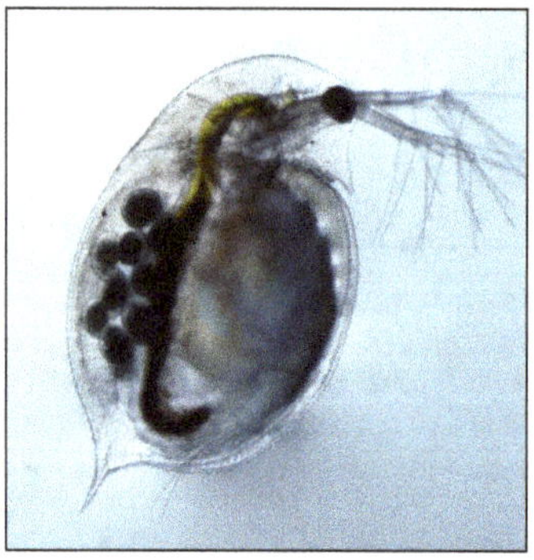

Daphnia pulicaria

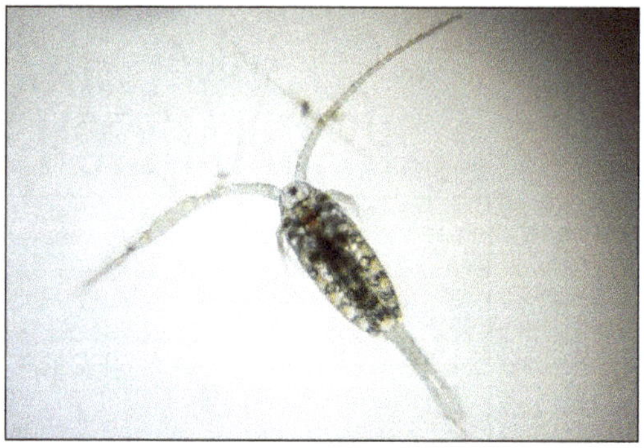

Diaptomus

Benthic organisms are major consumers and are also important recyclers of nutrients otherwise trapped in the sediments. Benthic organisms include invertebrates and bottom-feeding fish. Their feeding strategies vary widely. Some, such as clams, filter small bits of organic material from water as it flows by. Others eat detritus that has sunk to the bottom. The spread of the exotic invader, the zebra mussel, has caused dramatic changes in the water quality and ecology of Lake Erie in the past decade due to its high rates of filtration and high reproductive rate.

Not all organisms are easily classified as planktonic or benthic. For example, Chaoborus, Dipteran insect larvae, remain near the sediments in daytime and migrate to upper waters at night. These transparent predators ("phantom midges") migrate upward to feed on zooplankton, and are, themselves, a favorite food for fish. Mysid shrimp behave in a similar fashion and have been shown to migrate enormous distances (>100 meters) in Lake Tahoe each night.

The best known group of aquatic consumers is fish. Many small fish, such as sunfish and perch, primarily eat zooplankton. Tertiary consumers that prey on the smaller fish include larger fish and other carnivorous animals (loons, grebes, herons, and otters). Different species exploit different habitats (niches). Bass and pike are found in lakes that have beds of aquatic macrophytes suitable for spawning. Walleyes, on the other hand, spawn on a gravel bottom. Lake trout live only in very clear lakes with cold, well-oxygenated deep water. In contrast, carp are adapted to warm turbid, low oxygen lakes with mucky, high organic matter bottoms.

Decomposers

Decomposers, which include bacteria, fungi, and other microorganisms, are the other major group in the food web. They feed on the remains of all aquatic organisms and in so doing break down or decay organic matter, returning it to an inorganic state. Some of the decayed material is subsequently recycled as nutrients, such as phosphorus (in the form of phosphate, PO_4^{-3}) and nitrogen (in the form of ammonium, NH_4^+) which are readily available for new plant growth. Carbon is released largely as carbon dioxide that acts to lower the pH of bottom waters. In anoxic zones some carbon can be released as methane gas (CH_4). Methane gas causes the bubbles you may have observed in lake ice.

The decomposers can be found in all biological zones of a lake, although they are the dominant forms in the lower hypolimnion where there is an abundance of dead organic matter. Oxidation of organic matter by the decomposers (respiration) in the hypolimnion is responsible for the depletion of dissolved oxygen over the course of the summer, potentially leading to anoxic conditions (no dissolved oxygen). There is no source of oxygen in the hypolimnion to replace oxygen lost through decomposition. Stratification prevents atmospheric oxygen from being mixed deeper than the thermocline, and it is usually too dark for photosynthesis. Consequently, a large volume of organic matter from a variety of sources (e.g. wastewater, sinking algae, dying macrophytes, and organic sediment washed in from the watershed) leads to faster oxygen depletion and often complete removal of oxygen in the hypolimnion. The resulting anoxia has a profound effect on both the chemistry and the biology of the lake.

Trophic Status

Since the early part of the 20th century, lakes have been classified according to their trophic state. "Trophic" means nutrition or growth. A eutrophic ("well-nourished") lake has high nutrients and high plant growth. An oligotrophic lake has low nutrient concentrations and low plant growth. Mesotrophic lakes fall somewhere in between eutrophic and oligotrophic lakes. While lakes may be lumped into a few trophic classes, each lake has a unique constellation of attributes that contribute to its trophic status. Three main factors regulate the trophic state of a lake:

Rate of nutrient supply:

- Bedrock geology of the watershed;
- Soils;
- Vegetation;
- Human landuses and management.

Climate:

- Amount of sunlight;
- Temperature;
- Hydrology (precipitation + lake basin turnover time).

Shape of lake basin (morphometry):

- Depth (maximum and mean);
- Volume and surface area;
- Watershed to lake surface area ratio ($A_w : A_o$).

Trophic status is a useful means of classifying lakes and describing lake processes in terms of the productivity of the system. Basins with infertile soils release relatively little nitrogen and phosphorus leading to less productive lakes, classified as oligotrophic or mesotrophic. Watersheds with rich organic soils, or agricultural regions enriched with fertilizers, yield much higher nutrient loads, resulting in more productive, eutrophic (even hyper-eutrophic) lakes.

Eutrophication, the progress of a lake toward a eutrophic condition, is often discussed in terms of lake history. A typical lake is said to age from a young, oligotrophic lake to an older, eutrophic lake. Geological events, such as glaciation, created lakes in uneven land surfaces and depressions. The landscapes surrounding lakes were often infertile, and thus many lakes were oligotrophic. Eventually some of the shoreline and shallow areas supported colonizing organisms that decomposed unconsolidated materials into reasonably fertile sediments. Active biological communities developed and lake basins became shallower and more eutrophic as decaying plant and animal material accumulated on the bottom. Shallow lakes tend to be more productive than deep lakes, in part because they do not stratify, thereby allowing nutrients to remain in circulation and accessible to plants. They also tend to have a smaller lake volume, so nutrient loading from their watershed has a larger impact. There are undoubtedly exceptions to this typical progression from oligotrophy to eutrophy, where geology, topography, and lake morphology caused eutrophic conditions from the start.

This concept of lake aging has unfortunately been interpreted by some as an inevitable and irreversible process whereby a lake eventually "dies." In fact, many oligotrophic lakes have persisted as such since the last glaciation and some ultra-oligotrophic lakes, such as Lake Tahoe may have been unproductive for millions of years. Furthermore, research in paleolimnology has provided evidence that contradicts the idealized version of a lake becoming more and more eutrophic as it ages. Studies of sediment cores have suggested that the algal productivity of Minnesota lakes actually may have fluctuated a great deal during the past 12 - 14,000 years (the period since the last glaciation). Changes in climate and watershed vegetation seem to have both increased and decreased lake productivity over this period. Some lakes probably experienced high rates of photosynthesis fairly soon after glacial retreat and then became less productive until recent times. It is also possible that water sources for some lakes have changed over the past thousands of years through diversions of stream flow, for example. In such cases water supplies to a lake (and therefore nutrient supplies) could have changed, leading to changes in the lake's productivity.

However, lakes may be culturally eutrophied by accelerating their natural rate of nutrient inflow. This occurs through poor management of the watershed and introduction of human wastes through failing septic systems. Such changes may occur over periods of only decades and are reversible if anthropogenic nutrient loading can be controlled. In the 1960s this was a serious issue, exemplified by the hyper-eutrophic condition of Lake Erie. Although it was pronounced "dead," it eventually returned to less eutrophic conditions, when major point sources of phosphorus were controlled in the early 1970s (by spending millions of dollars to build advanced wastewater treatment plants).

In North America, most of the problems associated with the direct discharge of domestic wastewater have been successfully mitigated. Now the regulatory focus is on the much more difficult problem of controlling non-point sources (NPS) of nutrient pollution such as agricultural drainage, stormwater runoff, and inadequate on-site septic systems. NPS pollution is particularly difficult to address because it is diffuse, not attributable to a small number of polluters, and associated with fundamental changes in the landscape, such as agriculture, urbanization and shoreline development.

Eutrophication

Water Quality impacts Associated with Eutrophication:

- Noxious algae (scums, blue-greens, taste and odor, visual).
- Excessive macrophyte growth (loss of open water).
- Loss of clarity (secchi depth goes down).
- Possible loss of macrophytes (via light limitation by algae and periphyton).
- Low dissolved oxygen (loss of habitat for fish and fish food).
- Excessive organic matter production (smothering eggs and bugs).

- Blue-green algae inedible by some zooplankton (reduced food chain efficiency).
- "Toxic" gases (ammonia, H_2S) in bottom water (more loss of fish habitat).
- Possible toxins from some species of blue-green algae.
- Chemical treatment by lakeshore homeowners or managers may result (copper, diquat, 2,4-D, etc).
- Drinking water degradation from treatment disinfection byproducts.
- Carcinogens, such as chloroform (from increased organic matter reacting with disinfectants like chlorine).

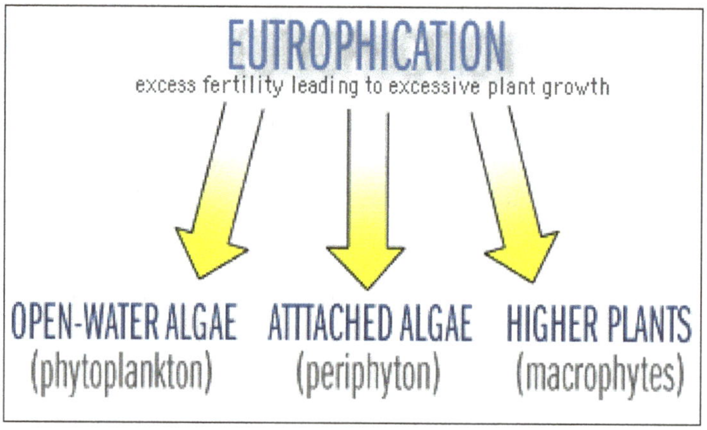

Biological Differences

Populations of algae and the animals that feed on them are lower in oligotrophic lakes because of low nutrient concentrations. Thus the water remains clear. Decay of the relatively small amount of organic matter in oligotrophic lakes does not completely deplete the hypolimnetic supply of dissolved oxygen. Therefore, lack of oxygen does not restrict animals from living in the hypolimnion of oligotrophic lakes. Lake trout, for example, require cold, well-oxygenated water and primarily live in the hypolimnion of oligotrophic lakes. Minnesota's oligotrophic lakes are found in the northeast region of the state, where infertile soils are covered with mixed conifer forests.

Extremely deep oligotrophic lakes such as Lake Superior and Lake Tahoe have hypolimnia that remain completely saturated with oxygen the entire year. However, many moderately deep lakes (with maximum depths greater than about 30 meters) may develop anoxia in the lower hypolimnion during late summer but may still be classified as oligotrophic because of their very low nutrient concentrations, low algal abundance, and relatively high transparency (high secchi depth). These lakes may have a two-story fishery, with warm and cool water fish in the epilimnion and metalimnion and cold water fish (such as trout) in the cold, oxygen rich portion of the hypolimnion. The cold-water fishery is therefore very sensitive to increased inputs of organic matter from sewage or erosion (external inputs), and to increased algal and macrophyte production (internal inputs) due to eutrophication since these factors will accelerate the rate and extent of hypolimnetic oxygen depletion in the summer.

Algae or macrophytes grow so thickly in some eutrophic lakes that light penetrates only a short distance and nutrients below that depth are not assimilated. As discussed earlier, phosphorus is typically the limiting nutrient in freshwater lakes, meaning that the plants deplete all available phosphorus before depleting other nutrients. In a hypereutrophic lake, algae may become so abundant that they suffer from self-shading. In those cases, photosynthesis is limited by light rather than by nutrients. When a great abundance of phosphorus is available in a lake, nitrogen may become limiting. In such lakes, certain species of blue-green algae that can fix atmospheric nitrogen have a clear competitive advantage and frequently become dominant. They dominate the algal community until another nutrient, or usually light, becomes limiting. In many infertile lakes in northeastern Minnesota, both phosphorus and nitrogen may be extremely low during midsummer. Since most sources of either point source or nonpoint-source pollution involve increased inputs of both N and P, these lakes are extremely sensitive to such pollution, irrespective of which is technically "most" deficient.

Eutrophic lakes show wide seasonal changes in their biological and chemical conditions. Because of the great amount of organic matter produced in these lakes, the decay rate is high in the hypolimnion, causing oxygen to be depleted. Therefore, eutrophic lakes frequently show a complete loss of dissolved oxygen below the thermocline during summers. Clearly, fish and most other animals cannot live in the hypolimnion of such lakes. Warm-water fish that can live in the epilimnion, however, can be quite productive. Bass, panfish, northern pike, walleye, carp, and bullheads thrive in many of Minnesota's eutrophic lakes. Complete or nearly complete oxygen depletion below the thermocline may also be a common feature of many moderately deep (10 to 30 m) mesotrophic lakes, if deep enough to stratify throughout the summer. Therefore, virtually complete anoxia below the thermocline does not necessarily mean that the lake is eutrophic.

Another oxygen-related problem in eutrophic lakes is winterkill. A dense snow cover over the ice reduces light penetration and keeps oxygen-producing photosynthesis from occurring. The high organic content of the water, however, provides considerable food for the decomposers. If the decomposers succeed in using all the available dissolved oxygen, a fish kill can occur.

In certain cases, a winterkill may lead to a more balanced fishery and possibly even improved water quality. Fish that survive a winterkill will have reduced competition for food for a period of time and so may grow faster and to a larger size. Fewer small fish reduces predation on the larger zooplankton, such as the water flea, Daphnia sp., leading to increased zooplankton grazing on algae and a resultant increase in water clarity. This general scheme, involving fishery manipulations to reduce the abundance of zooplanktivorous fish, has been termed biomanipulation, and is being tried in many urban lakes where it is economically impractical to reduce nutrient inputs enough to significantly reduce algae. In these situations the offending fish may be removed by intense stocking of gamefish, by intensive netting and trapping, or even by poisoning the entire fishery and starting over with greatly reduced planktivores.

Nutrients and their Role in Lake Ecology

Primary production in lake ecosystems depends on nutrients and light as essential resources. Phytoplankton take up nutrients dissolved in lake water; rooted macrophytes obtain nutrients from the sediments. Primary producers are potentially limited by carbon, nitrogen or phosphorus. Of these, carbon is the most common element in algal tissue and is also the most abundant in solution in lake water (CO_2, HCO_3^-, or CO_3^{2-}). Nitrogen (NO_3^-, NH_4^+, N_2) and phosphorus (PO_4^{3-}) are much less available, suggesting that phosphorus, followed by nitrogen, is most likely to limit algal production in lakes.

Limitation by other nutrients can occur: for example diatoms, algae characterized by hard cell walls containing silica, can be limited by silica availability. Light is also taken up and consumed like other algal resources. Light limitation can occur during algal blooms when cells close to the surface shade algae deeper in the water column, or when phytoplankton shade the lake bottom and prevent macrophyte growth.

Phosphorus

Phosphorus is essential in nucleic acids, phospholipids, adenosine triphosphate (ATP) etc., but is scarce in bioaccessible forms in lakes. Whereas carbon and nitrogen enter lakes by exchange of CO_2 and N_2 from the atmosphere, the only natural source of phosphorus is weathering from the watershed of PO_4^{3-} ions, which dissolve only poorly in water. Thus, bio-available phosphorus occurs in extremely low concentrations in freshwater and is taken up quickly by phytoplankton. In water containing dissolved oxygen, PO_4^{3-} can combine with iron to form insoluble salts that sink to the lake bottom, making phosphorus still less available to algae. Under anoxic conditions in the hypolimnion or the sediments, PO_4^{3-} becomes soluble again (a process called 'internal loading') and can stimulate algal growth through mixing processes.

The paramount importance of phosphorus as a resource for primary producers in lakes has been shown as a close relationship between phosphorus supply to lakes worldwide and the abundance of algae they contain, through studies of lakes polluted with phosphorus-containing human waste and laundry detergents, where phosphorus removal has eliminated noxious algal blooms, and through whole lake experiments in which enrichment with both phosphorus and nitrogen resulted in a bloom of cyanobacteria much higher than the growth of algae observed in a lake enriched only with nitrogen. The important role played by phosphorus in lake algal production is generally accepted today, and successful management of many lake ecosystems depends upon controlling phosphorus inputs.

Nitrogen

Algal cells require nitrogen to synthesize proteins. Typically nitrogen is available in lakes in higher concentrations than phosphorus. Most algae take up NH_4^+ ions (from decomposition) or NO_3^- ions (from bacterial nitrification of NH_4^+). However, many species of cyanobacteria can also use N_2 diffusing across the air–water boundary in a process called nitrogen fixation which forms NH_4^+. Lake ecosystems are not as frequently nitrogen limited as marine or terrestrial ecosystems. Lakes can only be nitrogen limited if the relative supply of phosphorus is greater than that of nitrogen. Adding wastewater from sewage to a lake can create this condition, which in turn favours the growth of nitrogen-fixing cyanobacteria because they have an alternative source of nitrogen. This is why lakes with excessive phosphorus inputs tend to have noxious blooms of cyanobacteria.

Formation of Lakes

Earth Movement

Tectonic Lakes: Due to the warping, sagging, bending and fracturing of the earth's crust, tectonic depressions occur. Such depressions give rise to lakes of immense sizes and depths.

They include Lake Titicaca, occupying a huge depression in the intermont plateau of the Andes, 12,500 feet above sea level the highest lake in the world; and the Caspian Sea, 143,550 square miles, the largest lake, almost 5 times larger than its nearest rival, Lake Superior.

Rift Valley Lakes: Due to faulting, a rift valley is formed by the sinking of land between two parallel faults, deep, narrow and elongated in character. Water collects in these troughs and their floors are often below sea level. The best known example is the East African Rift Valley which runs through Zambia, Malawi, Tanzania, Kenya and Ethiopia, and extends along the Red Sea to Israel and Jordan over a total distance of 3,000 miles.

It includes such lakes as Lakes Tanganyika (4,700 feet deep, the world's deepest lake), Malawi, Rudolf, Edward. Albert, as well as the Dead Sea 1,286 feet below mean sea level, the world's lowest lake.

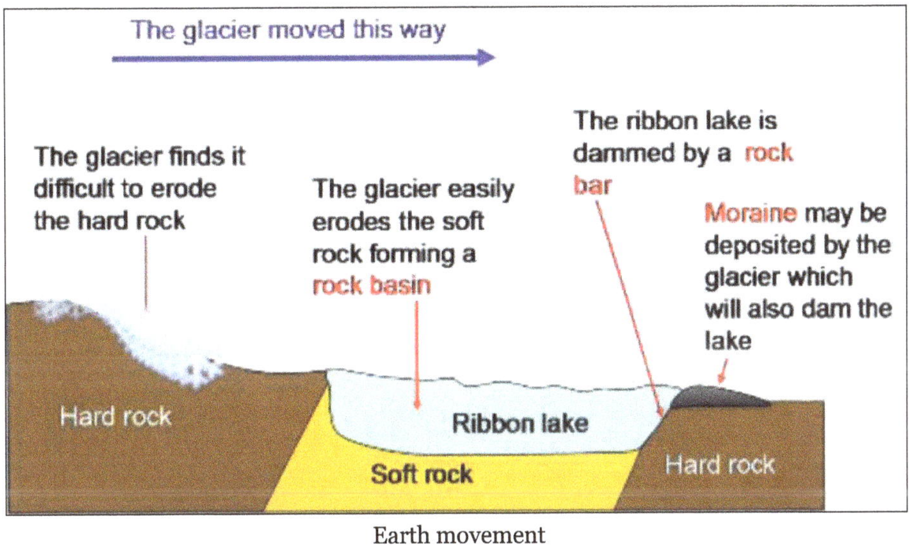

Earth movement

Glaciation

Cirque Lakes or Tarns: A glacier on its way down the valley leaves behind circular hollows in the heads of the valleys up in the mountains. Such hollows are the arm-chair-shaped cirques or corries. Their over-deepened floors may be filled with water to become cirque lakes e.g. Red Tarn in the English Lake District. Those that occupy glacial troughs are long and deep and are termed ribbon lakes, e.g. Lake Ullswater.

Kettle Lakes: These are depressions in the outwash plain left by the melting of masses of stagnant ice. They are irregular because of the uneven morainic surface and are never of any great size or depth e.g. the meres of Shropshire in England, and the kettle-lakes of Orkney in Scotland.

Rock-Hollow Lakes: These are formed by ice-scouring when valley glaciers or ice sheets scoop out hollows on the surface. Such lakes of glacial origin are abundant in Finland, indeed the Finns call their country Suomi — the Land of Lakes. It is said that there are over 35,000 glacial lakes in Finland.

Lakes Due to Morainic Damming of Valleys: Valley glaciers often deposit morainic debris across a valley so that lakes are formed when water accumulates behind the barrier. Both lateral and terminal moraines are capable of damming valleys e.g. Lake Windermere of the Lake District, England.

Lakes Due to the Deposition of Glacial Drifts: In glaciated lowlands with a predominant drumlin landscape, where drainage is poor, there are intervening depressions. These depressions are often water-logged, forming small lakes like those of County Down in Northern Ireland.

Volcanic Activity

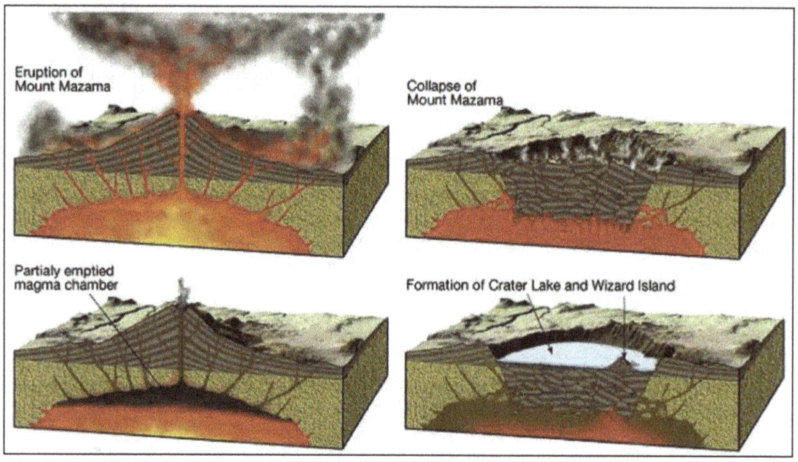

Volcanic activity

Crater and Caldera Lakes: During a volcanic explosion the top of the cone may be blown off leaving behind a natural hollow called a crater. This may be enlarged by subsidence into a caldera. These depressions are normally dry, bounded by steep cliffs and roughly circular in shape.

In dormant or extinct volcanoes, rain falls straight into the crater or caldera which has no superficial outlet and forms a crater or caldera lake. The outstanding ones are the Crater Lake in Oregon, U.S.A. which in fact occupies a caldera; Lake Toba in northern Sumatra and Lake Avernus near Naples.

Lava-Blocked Lakes: In volcanic regions a stream of lava may flow across a valley, become solidified and thus dam the river forming a lake, e.g. a lava flow blocks the Jordan valley forming the Sea of Galilee which is an inland lake, rather elongated in shape.

Lakes Due to Subsidence of a Volcanic Land Surface: The crust of a hollow lava flow may collapse. The subsidence leaves behind a wide and shallow depression in which a lake may form, e.g. Myvatn of Iceland.

Erosion

Karst Lakes: The solvent action of rain-water on limestone carves out solution hollows. When these become clogged with debris lakes may form in them. The collapse of limestone roofs of underground caverns may result in the exposure of long, narrow lakes that were once underground e.g. the Lac de Chaillexon in the Jura Mountains.

The large depressions called poljes, which normally do not have surface outlets, may contain lakes. During wet periods these may cover most of the polje floor but they shrink during dry periods due to seepage. An example is Lake Scutari in Yugoslavia.

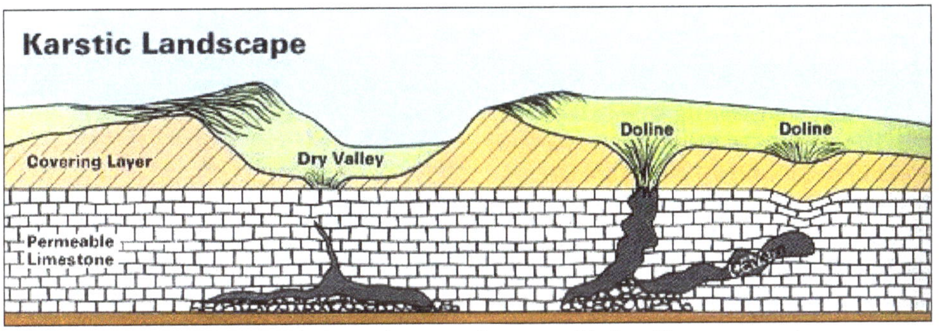

A Karst lake formed above the ground water table in a limestone region.

Solution is important in other rocks such as rock salt. Local subsidence may occur when the underlying beds of rock-salt are gradually removed in solution. Many of the meres of Cheshire, England, were probably caused by this, and are also the result of salt-mining operations.

Wind-Deflated Lakes: The deflating action of winds in deserts creates hollows. These may reach ground water which seeps out forming small, shallow lakes. Excessive evaporation causes these to become salt lakes and playas. These are found in the Qattara Depression in Egypt, and the Great Basin of Utah, U.S.A.

Deposition

Lakes Due to River Deposits: A river may shorten its course during a flood by cutting across its meandering loops, leaving behind a horseshoe-shaped channel as an ox-bow lake, e.g. those that occur on the flood-plains of Lower Mississippi, U.S.A. and Rio Grande, Mexico.

Lakes Due to Marine Deposits: The action of winds and waves may isolate lagoons along coasts by building spits or bars. As these lagoons of shallow water are enclosed only by a narrow spit of land, comprising mud, sand and shingle, they may drain away at low tide. They are commonly found off the deltas of large rivers such as the Nile and the Ganges.

In East Germany and Poland, lagoons are called haffs. Strong on-shore winds are capable of pushing coastal sand dunes landwards, and these may enclose marshy lagoons. This type of lagoon is well developed in the Landes of south-west France.

Lakes Due to Landslide Screes and Avalanches: Lakes formed by these processes are also known as barrier lakes. Landslides or screes may block valleys so that rivers are dammed. Such lakes are short-lived, because the loose fragments that pile across the valleys will soon give way under the pressure of water.

When they suddenly give way, the dammed water rushes down, causing floods. Examples of lakes of this type are Lake Gormire in Yorkshire, blocked by landslides; Ffynnon Freeh on Snowdon blocked by screes.

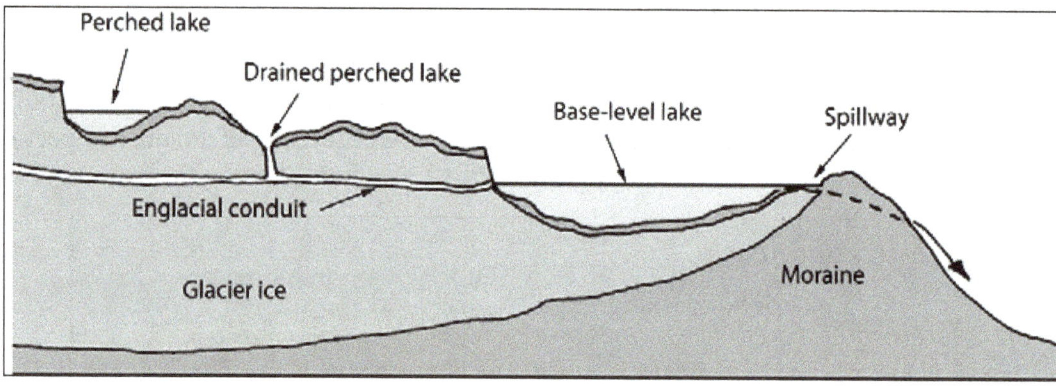

Deposition (a barrier lake formed by the deposition of an obstacle).

Human and Biological Activities

Man-Made Lakes: Besides the natural lakes, man has now created artificial lakes by erecting a concrete dam across a river valley so that the river water can be kept back to form reservoirs. Amongst such man-made lakes, the most imposing is Lake Mead above the Hoover Dam on the Colorado River, U.S.A.

Human activities (a lake made by constructing a concrete dam across a river valley).

Lakes Made by Animals: Animals like beavers are particularly interesting. They live in communities and construct dams across the rivers with timber. Such beaver dams are quite permanent and are found in North America, e.g. Beaver Lake in Yellowstone National Park, U.S.A.

Other Types of Lakes: These include ornamental lakes, especially made to attract tourists, e.g. Lake Gardens, Kuala Lumpur, Taiping Lakes. Man's mining activities, e.g. tin mining in West Malaysia, have created numerous lakes. Inland fish culture has necessitated the creation of many fishing-lakes.

Lake Zones

A typical lake has three distinct zones (limnetic, littoral and the benthic zone; Fig. 11) of biological communities linked to its physical structure. The littoral zone is the near shore area where sunlight penetrates all the way to the sediment and allows aquatic plants (macrophytes) to grow. The 1 % light level defines the euphotic zone of the lake, which is the layer from the surface to the depth where light levels become too low for photosynthesis. In most lakes, sunlit euphotic zone occurs within epilimnion. In transparent lakes, photosynthesis may occur well below the thermocline, into the perennially cold hypolimnion. The higher plants in littoral zone, in addition to being a food source and a substrate for algae and invertebrates, provide habitat for fish and other organisms different from the open waterzones.

The limnetic (pelagic) zone is the open water area where light does not penetrate to the bottom. The third component of the lake habitat is benthic zone (the bottom of the lake), covered by fine layers of mud in which animals live. In the littoral zone, there is enough light for rooted plants to grow, but beyond this zone, there are no rooted plants as the water is too deep for light to reach them. The deepest part of the open water forms profoundal zone, but this is relevant only in extremely deep lakes.

In each community, green plants produce food upon which all aquatic organisms depend. Hence they are called as autotrophs. Animals, which feed on these plants, are called as herbivores and those that eat other animals are called carnivores. Omnivores have a mixed diet and detrivores feed on dead and decayed plant and animal material. All the feeding habits link to form the food-web.

Fish are usually the largest animals in the lake, feeding on most invertebrates; some eat other fish and very few are herbivores. There are more than 20,000 species of fish out of which 7000 live in fresh waters. Fish feed on plankton, plants and invertebrates and are in turn eaten by carnivores such as birds, otters and man. At each step, food is eaten to gain energy and nutrients. Most of the energy generated is used to support daily activity and is dissipated into the environment as heat. Only surplus energy is available for growth and reproduction, such that the energy available for the next link in the food chain is always less than that which entered the present trophic level. Hence, there cannot be more carnivores than herbivores.

Pelagic Zone Community

Pelagic zone is the home of plankton and nekton. They are distinguished based on their swimming ability. Planktons are suspended in water whereas nektons are active swimmers. The phytoplankton consists of algae ranging in cells from 10 mm in diameter such as *Chlorella* to clusters of cells (*Volvox*). Several major groups of algae occurs the world over, although the species vary. Majority of the planktonic animals (zooplankton) are herbivores feeding on algae, but there are also carnivores. The most important species are crustaceans of two major groups: cladocera (eg. Daphnia) and copepoda. They range in size from 0.2mm to about 4mm long, with exceptional species reaching 1-2 cm. These animals are an important link between the algae and much larger creatures such as birds and fish. Fish are the most obvious predators of zooplankton. Most fish start feeding on rotifers and move on to cladocerans.

A most unusual community called neuston lives at the air-water interface of the lakes. Numerous types of algae and bacteria attach themselves to the surface, taking advantage of the surface tension. The cladoceran, Scaphalober is mucronata filter feeds on small particles while attached to the surface film. Typical pelagic fish include clupeids, which are an important link in the food chain of open water zones and a significant source of human food. Most birds feeding in this zone are fish eaters and either dive for their prey or catch it close to the surface. Birds (like geese and duck) are also attracted to these zones for refuge from terrestrial predators.

Benthic Zone Community

The term benthic applies to flora and fauna living at the bottom and those species that live by burrowing in it. Benthic organisms can live in the substrate (in mud and sand), move on the substrate surface, grow attached to the surface or move freely in the bottom. Benthic primary producers include cyanobacteria, all higher taxa of eukaryotic algae and flowering plants. Algae attached to a substrate are called as benthic, to distinguish them from phytoplankton, which live floating in water. In some shallow lakes, benthic algae may be an important source of food but most benthic animals beyond the littoral zone derive their food from detritus raining down from the open water.

Much of the bottom beyond the littoral zone will be covered with mud whose particle size and organic content depends on conditions specific to a lake. The larger inhabitants of this zone are mostly worms, larvae of chironomid flies, and molluscs. In addition there are numerous smaller animals like nematode worms and ostracods. Chironomids are very characteristic lakeside inhabitants, typical of many lake insects, with larvae living and growing in the water. The larvae are essential food for many fish and ducks. The phantom midges, Chaobrus spp, also have aquatic larvae whose adults emerge out from the water. They are interesting as they provide a link between benthos and pelagic zones. During the day, they rest in the substrate as members of the benthos and at night, they raise upwards in the water becoming pelagic to feed. They are voracious carnivores whose predation has a marked effect on other zooplankton. Many of the worms found in fresh waters have haemoglobin in the blood, which help trap oxygen, an advantage, as the benthic community suffers from lack of oxygen.

Molluscs are important components of the benthic fauna. The molluscs that are adapted to the benthic life are the burrowing bivalves, such as freshwater mussel. Mussels are important filter feeders and are usually the largest members of the lake community to feed directly on the smallest algae. The animals, which live in the deepest part of the lake water, are able to extract oxygen even from water with less oxygen content. At about 20-25% saturation, their respiration rates decrease, but they can survive periods of low oxygen concentration, although their feeding and growth are slower. Benthic animals have to tolerate long periods in summer, when not only the sediment but also the water lying on top is deoxygenated. Unlike pelagic fish, benthic fish cannot find their prey by sight, and many feel and taste for it with barbells (whiskers) around their mouth. Carps are typical benthic fish, which feed on whatever is available (molluscs, crustaceans, insects and plants), consuming more at higher temperatures.

Littoral Zone Community

The main groups of benthic invertebrates are also found in the littoral zone, but are often represented by different species. The inshore area is a more complex habitat, mainly due to the presence of macrophytes and supports many animal species. The only substrate for animals in this zone is rocks and the only food is algae either present in the water or attached to the rocks, and the detritus that is lodged among the stones. The advantage in this zone is that oxygen is never likely to be limited as water is constantly moving, dissolving more oxygen from the air.

Snails with a heavy shell and strong muscular foot are common, particularly if calcium content is high. Sponges encrust the rocks and leeches are common. Worms and freshwater shrimps are among the organisms that escape the waves by hiding. The common worms, larvae of caddis flies and chironomids, either attach themselves under the stones or burrow among the pebbles.

In sheltered places where sediments can accumulate, plants grow and make a tremendous difference to the number and type of animals found in this marginal area. Large plants in a lake are referred to as macrophytes, distinct from the other major plants, algae, most of which are microscopic. Most macrophytes need to be rooted in the bottom silt and therefore grow in relatively shallow water. Those at the lake edge normally grow tall and stick out of the water, referred to as emergent macrophytes. On gentle slopes, the most obvious macrophytes are tall aquatic grasses, notably reeds (Phragmites). The flat floating leaves of plants such as water lilies (Nymphaea) and pond weeds (Potamogeton) are rooted but have long stems. Further out in the lake are wholly submerged plants, some of which are rooted to the bottom, such as water milfol (Myriophyllum) and hornwort (Ceratophyllum) that float freely in the water. Among the plants that float on the surface in sheltered places, the smallest are duckweeds (Lemna), and the largest include water hyacinth (Eichhornia) and floating fern (Salvinia).

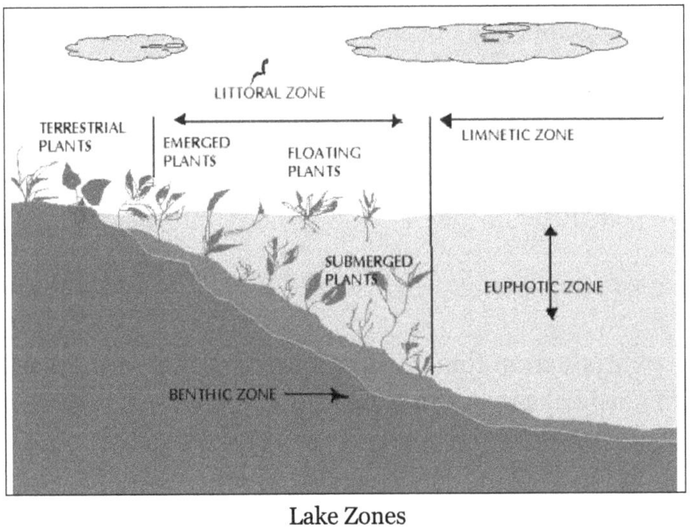

Lake Zones

The gradation of plants from land into water represents a transition from one environment to another and is known as zone of succession (hydrosere). As the emergent macrophytes die each year, their remains accumulate and trap debris and silt. Emergent macrophytes are similar to terrestrial plants except that they are always rooted in deoxygenated mud rather than well-aerated soil. Like other parts of the plant, the root cells need oxygen and if this is not obtainable from their immediate surroundings, it must be transferred down the stem from those parts of the plant, which are out in the air. For this purpose, roots, leaves and stems of macrophytes contain large air-filled spaces along which oxygen can travel to relieve the shortage (of oxygen) in the roots. Wholly submerged plants cannot obtain oxygen from the air and must extract all they need from water. They have very thin leaves and much of the oxygen released during photosynthesis goes to alleviate the shortage in the roots.

Animals in the littoral zone mostly live on the surface of leaves or stems or burrow among the plant roots. The littoral zone contains a complex mixture of plants, animals and microorganisms. Diatoms and blue-green algae are common components and the latter secretes a gelatinous layer, which attracts and holds other organisms. Pond snails are a conspicuous group of littoral zone animals along with many species of crustacea (shrimps and crabs) living among the plants. Littoral zone provides habitat for insects like dragonflies (Odonata), mayflies (Ephemeroptera), stoneflies (Plecoptera), caddis flies (Trichoptera) and midges (Diptera).

Coupling of Habitats

Benthic and pelagic habitats are not completely separate from one another as there are numerous connections between their components. Deepwater fauna is very dependant on the production in the euphotic zone. Animals can greatly accelerate the vertical transport of organic matter. For example, the faecas of zooplankton, especially copepod faecal pellets are an important source of food for the benthos. These faecal pellets contain a large amount of partially digested algae. Due to their large size, they sink to the bottom before the organic matter is decomposed.

Smaller lakes have more connections between the littoral and pelagic zones. Phytoplankton carried to the littoral zone provides food for mussels and cladocera. Some animals are littoral in one part of their life cycle and benthic in another such as some cyclopoid copepods, which are planktonic as juveniles and adults, but spend a part of their development as a resting stage in the sediments. During their development, fish change their habitat according to the location of the food. As juveniles, many cyprinids for example, eat plankton and therefore stay in the pelagic zone. When they become larger, they shift to feeding on the bottom during the day and move into the littoral regions of the lake at night. Some planktivorous fish spend the day near shore and move into open water at night to feed on pelagic plankton.

Productivity

Productivity is the rate of production for a given group of organisms, essentially representing the net growth rate of organisms. Lakes seldom suffer from shortage of water but are often unproductive due to lack of nutrients necessary for plant growth and reproduction.

Plants comprise the basic food material, directly or indirectly, for the whole lake ecosystem. Their abundance can therefore indicate the lake's productivity. Productivity is measured as the new carbon collected from the air and fixed as organic compounds by photosynthesis, so that it can be added to the total food supply in the lake. It is expressed as grams of carbon fixed per cubic metre ($g\ C/m^3$) per unit of time. The amount of small floating plants or phytoplankton indicates productivity, measured as the total amount of chlorophyll present. The amount of chlorophyll-a per cubic metre is an index of phytoplankton abundance and potential productivity.

Efficiency of carbon transfer within lake ecosystems is dependant on two factors: a) quantity of biomass created at primary producer level and b) composition of species, which determines efficiency of grazing and quality and quantity of fish (representing the highest trophic level in the system). After death, organisms at primary, secondary and tertiary levels decay and recycle nutrients.

Increased supply of nutrients due to anthropogenic activities (sewage, industrial effluents, fertiliser run-off from agricultural fields) changes the lake ecosystem causing eutrophication. Lakes with low supply of nutrients are called oligotrophic, while those with high nutrients are called eutrophic. Continuous eutrophication results from autotrophic production of internal organic matter by primary producers (photosynthetic plants and algae) from nutrients available in the lake, derived either from external sources (anthropogenic activity in the catchment) or internal recycling (decay of organic matter and dissolution from bottom sediments). In many shallow lakes, eutrophication may be manifested in macrophytic growth rather than phytoplankton. Efficient utilisation of nutrients depends on several factors, which define the growth condition, and hence the resultant biomass production at the primary producer level. Grazing of phytoplankton by zooplankton (secondary producers) and predation by fish (tertiary consumers) represent carbon transfer system of the lake.

Eutrophication can be highly detrimental to the quality of lake water and may limit its use. The detrimental effects may result from inefficient use of phytoplankton biomass that in turn is derived from high nutrient availability. This is mainly due to changes in dominance of algal species that are not consumed or inefficiently consumed by zooplanktons.

Lakes are classified based on productivity as follows:

- Oligotrophic lakes: These have low primary productivity, and low biomass associated with low concentrations of nitrogen and phosphorous (nutrients). They tend to be saturated with oxygen.
- Mesotrophic lakes: These are lakes in transition from oligotrophic to eutrophic conditions. Some depression of oxygen concentration occurs in hypolimnion during summer stratification.
    - Eutrophic lakes: These display high concentration of nutrients, high biomass productivity and low transparency. Oxygen concentrations can get very low (as low as 1 mg/L) in the hypolimnion during summer.
    - Hypereutrophic lakes: These are lakes at the extreme end of eutrophication with very high concentration of nutrients and associated biomass production. Anoxia or complete loss of oxygen takes place in the hypolimnion during summer.
- Dystrophic lakes: These are organic rich lakes (humic and fulvic acids) fed by external inputs of the lake (watershed).

Changes, which occur in the lake due to addition of nutrients (on account of human activities), are called as eutrophication. The additional nutrients (primarily nitrogen and phosphorous) most usually come from sewage, industrial effluents or agricultural fertilisers, or all the three.

Bacteria that use up the oxygen in the water first decompose raw sewage. Sewage always contains a higher concentration of phosphates than nitrogen due to increased use of phosphate containing detergents. In unpolluted waters, growth of algae is inhibited by the lack of phosphorous but the addition of these nutrients (through domestic sewage, other effluents etc.) increases algal growth in the receiving waters. While algae produce oxygen during the day due to photosynthesis, they use it up in the night. Thus, dense algal blooms cause suffocation of fish and other aquatic organisms. Increased algal production due to eutrophication initiates other changes in the aquatic ecosystem.

Detritus-dead Organic Matter and Detrital Dynamic Structure

Living biota constitute only a small portion of the organic matter of living systems, even in the most productive of freshwater habitats. Much of the organic matter in aquatic ecosystems is non-living and is called as detritus. The functioning of aquatic ecosystems depends on the cycling of organic carbon between living and non-living components. Detritus consists non-living organic matter in both dissolved and particulate forms. In aquatic ecosystems, organic compounds depend on dissolved organic carbon compounds (DOC) and particulate organic compounds (POC), the ratio usually being 6:1.

Organic matter originates from photosynthesis within the lake or the drainage basin. Organic loadings in particulate form are very rare and all organic carbon is transported in the dissolved form, which leads to stratified sedimentation, metabolised or exported. Very little of either the particulate or dissolved organic matter emanating from new photosynthetic production is consumed by animals or enters the consumer food chain, but enters dissolved and particulate detritus. These microbial and chemical fluxes are essential to the maintenance of nutrient recycling and metabolic stability of the ecosystem.

The amount of detrital organic matter is quite large and is chemically more recalcitrant to biological degradation than living organic matter. As a result, metabolism of detritus is slow but the quantity is large and much greater than that resulting from living organic matter ingested by animals. This slow degradation provides an inherent ecosystem stability that dampens brief volatile fluctuations of higher trophic-level populations, as they respond to these rapidly oscillating factors governing their growth and reproduction.

Seasonal Changes in the Lake Community

A lake's biological characteristics are determined largely by physical characteristics of the water column. Important physical characteristics include temperature, light transparency, as well as the total abundance of inorganic nutrients (largely a watershed characteristic). In addition, preceding populations influence successive populations by the assimilation of critical nutrients. As physical, chemical and biological conditions in the lake change over time, some species will be effectively eliminated from a lake those which cannot tolerate new conditions. Other species will be out-competed by organisms that are better adapted to the new environment. Some species flourish for a period of time and then give way to other species more compatible with changed conditions, such as warmer water, more daylight or lower concentrations of nitrogen and phosphorous. Short-lived plankton communities are characterised by these seasonal fluctuations; longer-living organisms such as fish, must be tolerant to different lake conditions throughout the year.

Phytoplankton community: Depending on the species, algae divide to form new individuals, two to ten times per month. It is possible for populations of algae in lakes to exhibit a succession of species during the course of one year. During stratification, the replacement of one dominant species by another may be determined by chemical factors or grazing effect of animals. Some species are present all the time but fluctuate in numbers, but many others show a clear seasonality and disappear for the remainder of the year. The time of appearance each year is characteristic to different species. In general, there is a tendency for diatoms to dominate in spring, followed by green algae and blue green algae later in the year. Generalisations are difficult to make because of the great variability observed among phytoplanktonic numbers and biomass from lake to lake. Several patterns are reasonably consistent, however:

- The seasonal periodicity of phytoplanktonic biomass is reasonably constant from year to year, if the freshwater ecosystem is not disturbed by outside influences, such as human modifications of the watershed, nutrient loading, etc.
- The extent of change in phytoplanktonic numbers and biomass through the seasons is usually very high, a thousand fold in temperate waters. Whereas seasonal variations in tropical waters are much lower, often as little as fivefold.
- Maxima and Minima of numbers and biomass of phytoplankton often are out of phase with rates of primary production. Primary productivity usually follows closely the annual cycle of incident solar radiation in temperate lakes.

Insects: Lake insects respond to seasonal cycles. Adults of different species emerge at different times of the year, such that there is seasonal succession of species. As the insects grow larger, they provide food for fish; the fact that each prey species reaches a maximum size at different times means that there is always food available for the predators that switch progressively from one species to the other.

Zooplankton: These populations also exhibit seasonal changes in their dominant species. Rotifers can increase their populations rapidly, and in strongly seasonal lakes are first to respond to increased algal populations in spring. Seasonal variations in flagellate and ciliate abundance are variable. Cryptomonad and other microflagellates are commonly numerically dominant across a spectrum of oligotrophic through hypereutrophic lakes. Generally ciliate abundance is coupled to maximum abundance of phytoplankton in the spring and early summer months. A secondary maximum in the autumn is frequently found in eutrophic and mesotrophic lakes.

Fish: Fish have a clearly defined breeding season and there is one period of the year when larval fish are abundant in the lake. In most lakes, where there is more than one species, their breeding may be separated in time so that the larvae of one species do not compete with those of another for the available food. In tropical lakes, some fish breed all through the year while others confine to breeding only during periods of flooding when they can breed in the shallow inshore areas or in the rivers which flow into the lake.

Energy Relationships of Freshwater Ecosystems

There are four main constituents of the environment alongside abiotic factors that form the freshwater ecosystem, which are as follows:

- Elements and Compounds of the ecosystem that are absorbed by organisms, required for food or respiration, many of which are also required by plants.

- Plants, which are autotrophic by nature, meaning they synthesize food by harnessing energy from the sun through photosynthesis. These plants (and some bacteria) are the primary producers, as they produce (and introduce) new energy into the ecosystem.
- Consumers, which are the organisms that feed on plant organic material. These may be primary consumers, which feed on plant material or secondary consumers, which feed on primary consumers.
- Decomposers attain their energy by breaking down dead organic material (detritus), releasing critical elements and compounds, which in turn are required by plants.

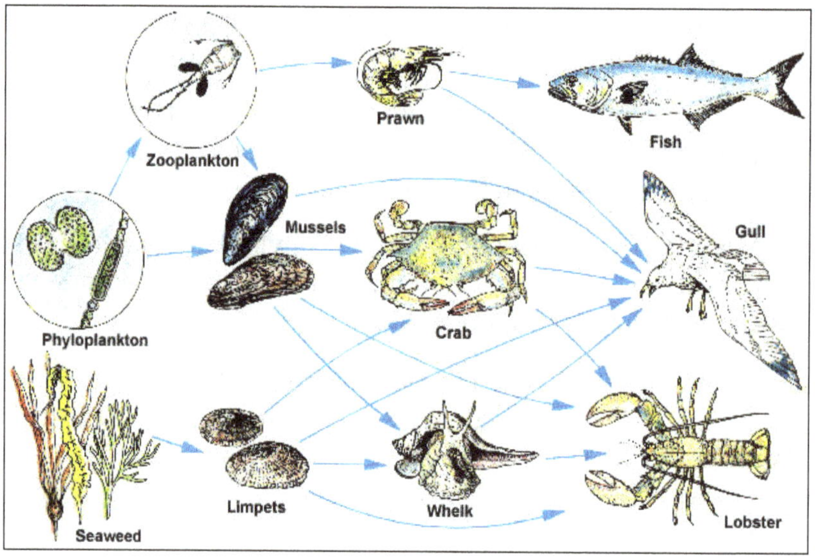

Food Web in a Typical Lake.

Biological communities within lakes may be organised conceptually into food chains and food web. The simplest illustration of the organisation of the organisms within an ecosystem is the ecological pyramid (food chain). The broad base of primary producers mainly supports overlying layers of herbivores, planktivores and less number of carnivores (predators). Food chains that allow a diversity of species to survive are divided into trophic levels, with plants providing the first trophic level, as they are the primary producers of most food chains. In almost all fresh water ecosystems, animals will be present, and form part of the many grazing food chains in the area. Other organisms leach the energy from dead organic matter forming detritus food chains. Such relationships allow the free flow of organic energy to be passed along from species to species, and provide an environment where food is available for them to survive. Food webs may be described in terms of both energy and nutrient (carbon, nitrogen or phosphorous) flows. The process typically begins with sunlight driven photosynthesis and respiration. Green plants capture energy from sunlight and convert it into living organic plant tissue. Lake photosynthesis includes algae and macrophytes, which together are the primary producers as they create organic material required by other organisms for nutrients and energy. Other fish feed on primary consumers and are called secondary consumers. They constitute the third trophic level. Still larger consumers such as large fish and man are tertiary consumers (fourth trophic level). Hence, energy and nutrients originating from the photosynthetic production of biomass and energy cascade through the food web. Respiration (the oxidation of organic material) releases energy that was originally captured from sunlight through photosynthesis. Decomposers are sinks for plant and animal wastes, but also recycle nutrients for photosynthesis. The amount of dead material in a lake far exceeds the living material. Detritus is the organic fraction of the dead material in a lake and is in the form of small fragments of plants, animals and microorganisms.

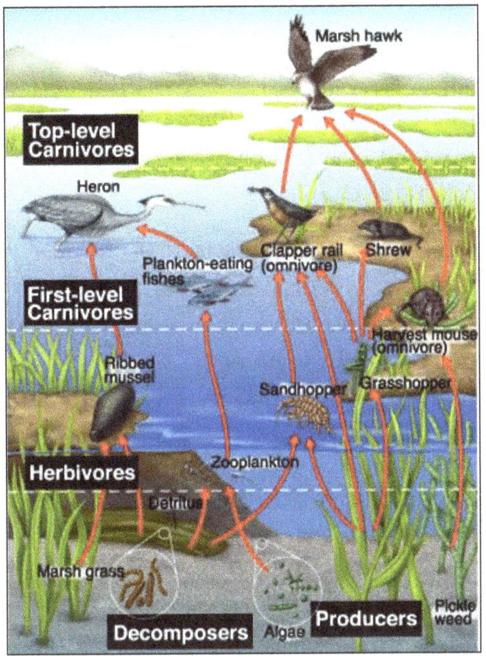

A Simplified Food and Energy Web in a Lake.

Figure, indicates an ecological pyramid where energy is passed along from autotrophic organisms to carnivores at the top of the chain. Those at the bottom of the food chain are usually the smallest (not always) and are almost inevitably the largest in number. Those which feed off these primary producers are less in number, usually because they are larger and require more than one primary producer as a means of food in order to survive. This situation continues to the top of the chain, where few secondary consumers are eaten by an even smaller amount of tertiary consumers.

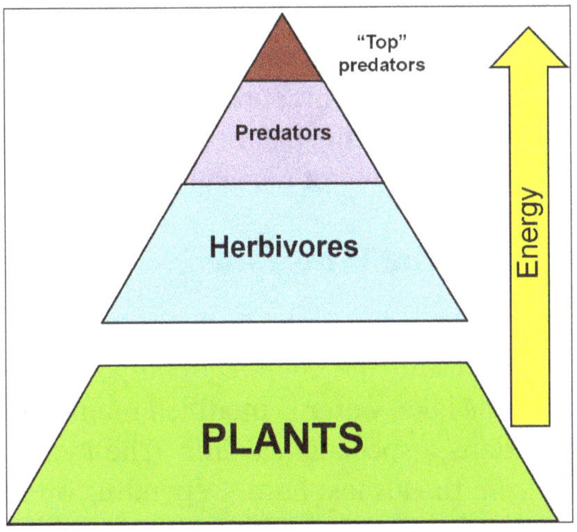

The Ecological Pyramid.

Oxygen released during photosynthesis adds to oxygen supplied by the atmosphere. In water layers where photosynthetic rates are high, such as during algal blooms, the water may become supersaturated, the saturation depending on the temperature of the water. Cold water can hold more oxygen than warm water. During periods of stratification, the only potential source of oxygen to the deeper lake is photosynthesis. The whole interaction of photosynthesis and respiration by plants, animals and microorganisms represents the food web. Food webs are usually complex and in many lake ecosystems, hundreds of different species can be involved. As the available energy decreases at each trophic level, a large base of plants (primary producers) is necessary to support few fish. These plants die and decompose or are eaten by primary consumers (second trophic level). The second trophic level involves zooplankton grazing on algae and eating fish, and invertebrates that feed on periphyton.

The most important fact to be understood here is that no matter what species occupies an area, chances are they will require another species to be in abundance in the area for feeding. The population of a particular species will depend on density dependant and density independent factors. Therefore the abundance of food in the ecosystem (a density independent factor) comes into play. The prey of a particular species will also require the existence of a food source, and other conditions conducive for them to survive.

Organisms within a food chain that can be assigned to the same position are said to occupy the same trophic level. Because ecological efficiency is always low, there is a rapid decrease in the productivity of subsequent trophic levels. Ecological efficiency refers to the ratio of predator productivity to prey production. This is the measure of energy passed from one link of the food chain to the next. Ecological efficiencies range from 5 to 20%, i.e., 80 – 95 % of the energy is lost at each transfer in the food chain. Because of these losses, energy must flow in one direction in the ecosystem. Trophic pyramids allow one to visualise the decreasing production with increasing height of the trophic level. The blocks that represent the production rates become smaller as one goes up to the higher trophic levels, giving rise to the appearance of a pyramid.

Cycling of Matter

Transfer of energy in ecosystems is always coupled with transfer of matter. The most important difference is that matter can circulate in an ecosystem whereas energy can only flow through the ecosystem. The biochemical cycles occurring in the lake are not closed. Lakes receive substances from the watershed and atmosphere; they export substances via outflow and atmosphere and bury substances in the sediments. Depending on the balance of these processes, water passing through the lake can be either enriched or diminished of particular compounds.

Modifications of the Water in the Watershed

Before entering the lake, water can be chemically altered, while passing through the lake watershed. The geology of the watershed and the chemical composition of the precipitation are extremely important. On its way to the lake, water is modified mainly by rock weathering. This leads to an enrichment of bivalent cations, especially calcium. The increased content of calcium permits uptake of carbon dioxide from the atmosphere. Depending on the pH, CO_2 is converted into bicarbonate and carbonate. Recycled matter from dead organisms also dominates nutrient loading in fresh waters. There are fundamental differences in the geological origin of different nutrients, such as nitrogen and phosphorous. Nowadays, much of the nitrogen and phosphorous entering the lakes is derived from soils. Human activities can greatly accelerate the release of nutrients into the lakes and streams. Non-point source pollution due to agricultural activities tend to increase the N:P ratio in the water bodies, while point source pollution due to domestic sewage (being rich in P), lowers the ratio of N:P.

Modifications of the Water Within Lakes

Water from the watershed enters a lake from surface and through groundwater inflows. The retention of a particular element may be calculated as the difference between the total amount of the element entering the lake via the surface and groundwater inflows minus the total amount leaving the lake provided the exchange with the atmosphere is negligible. If the retention capability of the sediment exceeds its threshold value, then the sediment releases substances instead of trapping them, causing a negative retention. However, negative retention is possible only for a short time, for those substances that have been previously deposited can be released from the sediments. In the long run, the sediments act as a sink for substances in the lake.

Balances are more complicated for elements that are exchanged with the atmosphere, like carbon and nitrogen. Exchange of gases with the atmosphere tends to reduce the degree of under-saturation and super-saturation in the surface waters. The net balance of exchange depends on biological processes in the lake. A lake, heavily loaded with organic matter and high rates of respiration by heterotrophic bacteria may have a net export of CO_2. The total balance for nitrogen in the lake depends on the balance between nitrogen fixation and denitrification. Nitrogen fixation lowers the concentration of nitrogen in water, allowing for an influx of this gas from atmosphere. Denitrification increases nitrogen in the water and may result in export from the surface water to the atmosphere.

Carbon Cycle

Dissolved Inorganic Carbon (DIC), Dissolved Organic Carbon (DOC) and Particulate Organic Carbon (POC) are the most important pools of carbon in aquatic ecosystems. DIC consists of carbon dioxide, bicarbonate, and carbonate and is indirectly affected by pH, and photosynthetic and respiratory activity of the aquatic organisms. The most important input of DIC is from atmospheric CO_2 and the dominant abiotic output is release of dissolved carbon dioxide into the atmosphere. Respiration is the biological input and photosynthesis is the biological output. In addition to allochthonous inputs, the most important sources of DOC are secretion and excretion by organisms from all trophic levels and autolysis of detritus. Uptake by heterotrophic microorganisms especially bacteria is the most important source of DOC. Excretion products are rapidly utilised by bacteria. POC consists of carbon bound in organisms and detritus. Primary production is the source of POC. There is shift in the POC pool due to organism death, feeding, parasitism, etc. POC can be transformed into DIC by respiration and into DOC via secretion, excretion and autolysis. POC is lost from the pelagic region and imported into the benthic zone through sedimentation.

Nitrogen Cycle

The most abundant nitrogen component in lakes is dissolved elemental nitrogen; however, only few organisms can utilise this form. Nitrogen fixation requires the enzyme nitrogenase that is found only in prokaryotes. For autotrophic organisms that do not produce nitrogenase, dissolved nitrite, nitrate and ammonium are the most important inorganic nitrogen forms. These forms of nitrogen are carried into the lakes through surface water, ground water and precipitation. Autotrophic organisms can use all three forms. But the decomposition of N-rich organic compounds releases only ammonium. In contrast to vertebrates that excrete either urea or uric acid, zooplankton excrete ammonium. The nitrate respiration that takes place in anoxic conditions transforms nitrate to either ammonium (nitrification) or elemental nitrogen (denitrification). Nitrification, which takes place under aerobic conditions, transforms ammonium into nitrate (through the intermediate formation of nitrite). These three forms of nitrogen are taken up by phytoplankton, and hence have a minimum concentration at locations where photosynthetic activities are maximum. Dissolved organic nitrogen comes from the excretion of organisms and decomposition of detritus. This consists of polypeptides and other amino acids. Simple amino acids are taken by bacteria and are therefore present in low concentrations. Microorganisms also excrete peptidases that assist in the splitting of peptides into smaller units.

Phosphorous Cycle

Phosphorous occurs only as phosphates in ecosystems. It is bound to organic compounds by ester bonds between phosphoric acid derivatives and carbon chains. Orthophosphates is much less mobile than nitrogen in the sediments and precipitates easily as it combines with several cations to form highly insoluble products. The most important material sink of soluble reactive phosphorous (SRP) is uptake by algae and bacteria. The carbon heterotrophic bacteria can be P-autotrophic i.e., in taking up the inorganic phosphorous they compete with phytoplankton for soluble reactive phosphorous. Despite the fact that phosphorous is often the growth-limiting factor, algae and bacteria lose soluble reactive phosphorous and to a lesser degree, low molecular weight non-reactive phosphorous. During the growth period, soluble reactive phosphorous in the epilimnion may be reduced to its limit of detection. Periods in which zooplankton grazing exceeds phytoplankton production can usually be recognised by an increase in SRP concentration. SRP and particulate phosphorous have complementary seasonal changes.

Lake Benefits, Disturbance and Restoration

Lakes occupy such a small fraction of the landscape belies their importance as environmental systems and resources for human use. They have intrinsic ecological and environmental values. Besides, human use, they are used for many commercial purposes including fishing, transportation, irrigation and industrial water supply, and function as receiving waters for wastewater effluents.

They moderate temperatures and affect the climate of the surrounding area. By storing water they help regulate stream flow, recharge ground water aquifers and moderate droughts. They provide habitat to aquatic and semi aquatic plants and animals, which in turn provide food to many terrestrial animals, adding to the diversity of the landscape.

The myriad ways in which humans use lakes, along with numerous pollutants generating activities, have stressed lake ecosystems in diverse ways, frequently causing impairment of lake quality for other uses. Stresses to lakes arise from easily identifiable point sources such as municipal and industrial wastewater, non-point degradation like urban and agricultural run-off within a lake's watershed, and the most insidious long-range atmospheric transport of contaminants. Major degrading factors include excessive eutrophication due to nutrient and organic matter loading; siltation due to inadequate erosion control in agricultural, construction, logging and mining activities; introduction of exotic species; acidification from atmospheric sources and acid mine drainage; and contamination by toxic (or potentially toxic) metals such as mercury and organic compounds such as poly-chlorinated biphenyls (PCBs) and pesticides. In addition, physical changes at the land-lake interface (eg. draining of riparian wetlands) and hydrologic manipulations (eg. Damming outlets to stabilise water levels) also have major impacts on the structure and functioning of these ecosystems.

Anthropogenic Disturbances of Lakes

Eutrophication: Causes and Consequences

Eutrophication has proved to be one of the most widespread and serious anthropogenic disturbances to aquatic ecosystems. The major cause for eutrophication is the increased loading of nutrients, especially phosphorous (P). Increasing wastewaters, introduction of phosphorous containing detergents, use of fertilisers, and erosion in the watershed are the major reasons for increased loading of nutrients. Fertilisers are the major source of nitrogen, while phosphorous in the fertiliser may be immobilised in the soil. Flooding can however transport large quantities of phosphorous due to leaching from the soil. Domestic sewage primarily results in increased P loading, containing on the average, an N:P ratio of 4:1 leading to a shift from phosphorous to nitrogen. An undesirable consequence of eutrophication is the development of cyanobacteria (blue-green algae) that tend to form dense "surface blooms", excreting organic compounds that impart bad odour and taste, creating serious problems in drinking waters. Some of the problem-causing algae are Anabena, Aphanizomenon, Microcystis, Limnothrix and Planktothrix.

The main effect of eutrophication in the hypolimnion and sediments is the increased consumption of oxygen. Anoxic conditions at the sediment surface and hypolimnion, impoverish the deep water fauna. This condition may also lead to a series of chemical and microbial processes like nitrate ammonification, denitrification, desulphurication and methane formation. Even pelagic fish, which release their eggs in the open water that sink to the bottom, cannot continue to reproduce normally in a lake with anoxic sediment surface. The release of phosphorous from the sediments is extremely important as it accelerates eutrophication. Fish kills may result from advanced eutrophication, initiated by elevated pH as a result of high photosynthetic rates and total ammonium concentration. Increased pH causes a shift from non-toxic ammonium ion to toxic ammonia.

The most important effect of high pH on aquatic animals is related to changes from non-toxic ammonium ion to toxic ammonia (as the disassociation of ammonia is controlled by pH). For $pH < 8$ there is only ammonium ions, which become ammonia at elevated pH (>8.5). High ammonium concentration and pH, caused by photosynthesis in poorly buffered and nutrient-enriched lakes, can lead to fish-kills.

Biodiversity of Inland Waters

Very little is known about the biodiversity of most microbial, plant and animal groups. Rate of loss of biodiversity varies both because of differences in intensities of disturbances to the habitats as well as the number of species inhabiting particular areas.

Water quality alterations as disturbances to biodiversity: The fresh waters of the world are experiencing accelerating rates of degradation. These major sources of disturbance impact the biodiversity of fresh waters in many ways. Direct release of chemical toxicants into the surface waters is very common. Many forms of heavy metals, inorganic nutrients and organic compounds enter the aquatic environment. Although some are inactivated by chemical precipitation or oxidation, many have long resident times. Despite dilution and dispersion in the aquatic environment, biomagnification of both metals and organic compounds is common, which increases toxicity exponentially. The biological effects of these compounds are still unknown.

Plant nutrients, particularly phosphorous and nitrogen, can lead to increased organic productivity of fresh waters. This eutrophication rate leads to enhanced rates of decomposition and in chemical conditions greatly reduce or eliminate suitable habitat for many species of plants and animals. Excessive loading of organic matter results from organic sewage from human populations, industrial effluents and agricultural runoff. Another factor that contributes to reduction in habitat availability is the suspension of finely divided organic and inert inorganic matter. Erosion and transport of such substances are increasing as a large amount of vegetation between land and water is eliminated, primarily for agricultural expansions.

Changes in biodiversity in freshwater ecosystems can arise from many disturbances. The introduction of certain competitively superior species can result in marked losses of biodiversity. Dense-floating macrophyte communities and other eutrophication associated excessive plant productivity often result in deoxygenation and reduction in habitat and elimination of plant and animal communities. Effective utilisation of freshwater resources is complicated by distribution of humans and their exploitation of regions of low water availability. Accelerating human induced changes in hydrological patterns associated with flood controls and water supply changes further obscure efficient utilisation practices. Control and reversal of degradation requires a proper economic valuation of freshwaters for efficient, conserved utilisation of water supplies for agricultural, industrial and domestic purposes.

Shifting resilience, stability and biodiversity of freshwater ecosystems: A central feature of stability is the ability of a property (nutrient concentration, population and community) of an ecosystem to return to a steady-state equilibrium following a disturbance. Resilience or relative stability is the measure of the rate at which the property or system approaches steady state following disturbance. Resilience or relative stability in an aquatic ecosystem is governed by process rates, that is, the rates at which energy and materials are transferred among biotic components. Because much of the organic matter is detrital, most of the assimilation and decomposition metabolism is microbial (bacterial, fungal etc.). Each of the species constituting the biotic groups has a physiological range for growth, competition and reproduction within the environmental constraints. Some of these ranges differ with spatial or temporal changes in life histories of different species, which can affect species responses to disturbances.

Physical and chemical environmental conditions mediated by a host of biogeochemical/climatic/meteorological factors regulate the bounds of nutrients and the energy flux rates. Major deviations from natural ranges arise from human or abnormal natural disturbances. The integrated ecosystem responds to major disturbances by shifting dominant metabolism from one species to alternative species as the physiological tolerance of certain species crosses the threshold limit resulting in their death. Greater biodiversity may have a greater collective effect by improving the ability for ecosystems to recover from large disturbances. Reduced biodiversity increases vulnerability by reducing the total collective physiological tolerance of the community to large habitat changes/ disturbances.

Extent and timing of disturbances and shifting stability: The differences in plant productivity result in different types of chemical composition of organic matter loaded to the lakes. Because of the large amounts and relative recalcitrance of dissolved and particulate organic detrital sources from higher plants, heterotrophic utilisation of organic matter is slowed. Alterations to the quality and quantity of dissolved organic matter entering the aquatic ecosystem from the drainage basins will influence this inherently chemically mediated metabolic stability. In this context, forested watersheds, undisturbed riparian floodplains and wetland/littoral zones are important to the metabolic stability within lake ecosystems.

A common disturbance to lake ecosystem involves progressive increases in loading of nutrients. This increase, particularly among shallow lakes that predominate globally, a marked shift in productivity occurs from the pelagic to surfaces (attached surfaces) associated with living aquatic plants and the particulate detritus of dying (senescing) macrophyte biomass.

Biodiversity and nutrient recycling: Greater biodiversity results in greater physiological diversity to cope with natural vagaries in environmental parameters, intensifying competition for resources (like nutrients). Nutrients once acquired are intensively recycled among the attached biota and conserved. Nutrient retention is very high within micro communities and collectively, the attached habitats. The increased dependence on labile organic substrates from algae reduces resilience to disturbances.

Predation-induced shifts in energy fluxes and biodiversity: In certain moderately productive lakes, piscivore consumption of planktivorous fishes can lead to sporadically enhanced development of zooplankton. The higher grazing rates of zooplankton can result in selective reduction of larger algae. Nutrient recycling also increases particularly as decomposition of these algae and metabolically coupled bacteria is accelerated by micro consumers and tightly retained in the biota. Here, the biodiversity among micro biota would increase and compensate for loss of larger forms by zooplankton grazing.

Bioindicators - Macroinvertebrates and Fish

Bioindicators are organisms whose presence, absence or condition provides information about the environmental quality. Every organism has unique environmental requirements to be healthy and reproduce successfully. The presence or absence of healthy populations of organisms within their habitats is a sign of unique environmental characteristics. The advantage of using bioindicators over chemical and physical tests to evaluate water quality is that the presence of living organisms inherently provides information about water quality over time. Chemical and physical tests give information that is accurate only for that moment when sample is taken. The presence of a mixed population of healthy aquatic insects or fish usually indicates that the water quality has been good for some time. The absence of bioindicators at a site that appears good according to chemical and physical sampling might prompt further investigation for toxic contaminants or periodic degradation of water quality.

In lakes, benthic macroinvertebrates are taken into consideration to get an idea of water quality. Benthic macroinvertebrates include aquatic insects, worms, shellfish, crustaceans and other animals without backbones that are large enough to be seen without a microscope and live at the bottom of a water body. Many species of mayfly nymphs, caddisfly larvae and stonefly larvae, for example, can survive only in swift, cool, and oxygenated water. Their presence at a sampling site is generally a sign of good water quality. Black fly larvae, midges, leeches and aquatic worms on the other hand, are quite tolerant of pollution. They can be found in waters of both good and poor quality. If they are the only types of macroinvertebrates found at a site, chances are the site is silty and has low dissolved oxygen. Such conditions might represent a polluted lake. Apart from using the presence or absence of certain indicator species to determine water quality, the number of different species can provide more information. The three most pollution intolerant orders of aquatic insects are Ephemeroptera (mayflies), Plecoptera (stoneflies) and Trichoptera (caddisflies). The higher the percentage of these pollution intolerant species in relation to the percentage of tolerant species at a site, the better the water quality.

The health of resident fish species will be indicative of overall water quality. The condition of the lake is determined by comparing the length of the fish to its weight. Heavier the fish for its length, better the condition.

Freshwater mussels, like aquatic insects, serve as bioindicators. Each species of mussel has different environmental requirements. Some species like Elliptio complanata, the Eastern Elliptio Mussel, are more pollution tolerant than species like Margaretifera margaretifera, the Pearl shell Mussel. Freshwater mussels are filter feeders that use a siphon to pump water into their shell where they use gills to simultaneously retrieve oxygen and extract food from the flowing water. Because they move so slowly, mussels must be able to obtain their food and oxygen from one spot. If their habitat becomes inundated with pollution such as toxins or silt, some of them will most likely perish.

Perhaps the most interesting aspect of freshwater mussel natural history is reproduction. Larval mussels, called glochidia attach to the gills or scales of fish. The glochidia develop for a while attached to the fish and eventually drop back to the bottom. The glochidia do not seem to harm the fish. Scientists think that the mussel glochidia will only survive attached, in some cases, to particular species of host fish. For example, the Pearl shell Mussel mentioned above usually uses salmon and trout as a host. If obstacles such as dams, poor water quality or over fishing interfere with the survival of the fish, they are likely to interfere with the survival of the mussels too. Because freshwater mussels can live for decades, their presence or absence can provide even more information about the history of water quality at a site.

Restoration of aquatic ecosystems: Lakes are more susceptible to function as sinks for pollutants, which means that pollutants tend to accumulate and potentially increase the toxicity with time. The need for management of lakes and rivers reflects an inability of the ecosystems to operate in self-sustaining ways because of the interference or damage that exceed the capacities of the ecosystem for self-restoration. Most demands for management are a result of disturbances by human activities. Four major types of degradation of surface waters have occurred in recent times:

- Nutrient loading, particularly phosphorous and nitrogen, have resulted in nutrient enrichment and excessive growth of algae and macrophytic plants. This accelerated eutrophication has led to evolving suitable remedial measures to mitigate the problems and return the lake to some state of lower productivity.

- Soil erosion, primarily from removal of vegetation cover and agricultural activities, leading to siltation, loss of volume in the aquatic habitat, etc.
- Excessive loading of hydrogen ions associated primarily with strong acids. These acids result from gases from fossil fuel combustion products that are dissolved in precipitation or adsorbed to particles that are deposited from the atmosphere onto surface waters. The resulting acidification of poorly buffered waters not only increases acidity and alters the osmoregulatory capacities of organisms but also alters the solubility and bioavailability of many metals and other ions of variable toxicity.
- Introduction of toxic materials such as heavy metals, chlorinated hydrocarbons and radioactive materials, the sources of which are often diffuse (non-point) sources, making their control very difficult.

Lake Management and Restoration

Lake management and restoration are focussed mainly on excessive nutrient loading and land management.

Lake restoration efforts have therefore been directed towards reducing the loading of phosphorous and to some extent nitrogen to the surface waters by advanced wastewater treatment, diversion and land management or reducing the phosphorous load in wastewaters by reducing the phosphorous content in detergents. Prevention of nutrient pollution and eutrophication is a long-term solution. Reductions in the nutrient loading require evaluation of diffuse and point sources within the drainage basin and a systematic program of reduction and control.

Nutrient Control

Some of the common methods that are used singly or in combination to reduce the nutrient availability are listed below:

- Nutrient removal by advanced treatment and land management: Regulation and reduction in external loadings of nutrients to eutrophic lake is the best of long-term corrective measures. Advanced wastewater treatment is practical only where used water is collected and phosphorous removed to quantities that will not alter the lake ecosystem. Agricultural land management practices including manure storage and changes in tillage can also reduce nutrient loading.
- Nutrient diversion: Occasionally diversion of the major nutrient loadings is adequate to restore a eutrophic lake. In situations where the primary source of nutrient loadings is defined, wastewater or storm water may be diverted without appreciable alteration of lake hydrology. Diversion is most likely to be successful in rapidly flushed lakes, where internal nutrient loading from sediments is small.

Dilution and Flushing

If a major source of nutrient-poor water is available, addition of such water to an enriched lake may be adequate to dilute nutrient levels to suppress algal productivity. If sufficiently large quantities can be added, the water alone can be adequate to flush out algal cells and maintain a lower productivity by dilution. Lower algal biomass can be maintained even with nutrient rich water if flow is adequate to wash out the algae faster than algal growth rates. These methods are used in conjunction with induced lake circulation, chemical precipitation and biological manipulations.

Phosphorous Precipitation and Inactivation

Recovery from eutrophication in which essential nutrients are reduced to low concentrations can be a slow process if lake water renewal times are long and nutrient releases from the sediments to the epilimnion are large. Phosphorous is precipitated by clay and carbonate particles present in water. Aluminium also precipitates phosphorous. Aluminium sulphate (alum) or sodium aluminate forms a precipitate of aluminium phosphate or a colloid of aluminium hydroxide, depending on the pH and alkalinity. In both cases, phosphorous is bound and transported to the sediments without aluminium toxicity if neutral conditions are maintained. Further addition of aluminium to the sediments retards phosphorous migration to the overlying water.

Sediment Oxidation

High bacterial decomposition of organic matter in sediments results in anaerobic conditions. These reducing conditions lead to reduced iron and release of associated phosphates into interstitial sediments and overlying water. Additions of nitrate as an alternative electron acceptor for oxygen delays the reduction of iron and release of phosphate.

Aeration

Two methods have been employed to aerate water of stratified lakes, artificial circulation and hypolimnetic aeration. Oxygenation can induce precipitation of phosphorous, iron and manganese and lower ammonium content and increase pH. Artificial water circulation by pumps, jets, bubbled air or other gases, destratifies the water strata, which reduces the competitive advantages of cyanobacteria. Internal loading of phosphorous often declines from renewed aerobic conditions at the sediment-water interface if adequate iron is present. Otherwise iron has to be added.

Sediment Removal

Shallow eutrophic lakes can often be restored by removal of sediments. This technique simultaneously removes large quantities of nutrients stored in the sediments, increasing the mean depth of the basin and removing toxic substances.

Food-web Manipulations

The management of food web structure by manipulation of animal herbivory, predation and nutrient recycling can influence the qualitative and quantitative composition of algae on a seasonal basis. Nutrient concentrations and loading rates set limits on the types of phytoplankton and their productivities. Changes in zooplanktonic herbivore types and densities can alter algal biomass and species composition below those anticipated by the nutrient loadings to the lake ecosystem. The phytoplankton productivity is usually rapidly replaced by smaller algal species that may be less obvious for human uses of the water bodies.

Control of Macrophyte Biomass

Excessive growth of macrophytes, particularly certain exotic species like water hyacinth (Eichhornia crassipes), hydrilla and duckweeds, can curtail or eliminate the use of lakes.

Draw down of water level: The drawdown of water level is a multipurpose restoration technique for the management of aquatic plants in lakes and to modify the habitat for fish populations.

- Mechanical removal: Aquatic macrophyte communities can be controlled by numerous manual and mechanical methods. Harvested plants are preferably removed from water, simultaneously removing appreciable quantities of nutrients.
- Biological controls: Introduction of phytophagous insects and fish or plant pathogens such as fungi and viruses has been used as species-specific control agents to reduce the biomass of macrophytes.
- Chemical controls: A number of general and moderately selective herbicides have been effectively applied to control aquatic macrophytes. Such compounds are used in combination with other control methods.

The term restoration means the reestablishment of predisturbance aquatic functions and related physical, chemical and biological characteristics. The objective is to emulate a natural, self-regulating system that is integrated ecologically with the landscape in which it occurs. Often, restoration requires one or more of the following processes: reconstruction of physical conditions, chemical adjustment of the soil and water; and biological manipulation, including the reintroduction of absent native flora and fauna.

The conservation and protection involves not only buffering wetlands from direct human pressures, but also maintaining important natural processes that operate on wetlands from outside, which may be altered by human activities. Management towards this end should emphasize the long term sustenance of historical, natural wetland functions and values. Restoration is thus a good opportunity to manage wetlands for broad wildlife goals, as restored wetlands provide enhanced wildlife benefits, in addition to other benefits, concurrently.

The preliminary step that has to be implemented in restoring lakes for their long-term sustenance includes:

- Pollution impediment: Wastewater, solid and semi-solid wastes entering in to the lake from external sources must be stopped before any restoration work is implemented.
- Harvesting of Macrophytes: Water hyacinth and other nuisance vegetation present in the lake, causing eutrophication, must be removed manually or mechanically. Weed infestation can also be controlled by applying chemicals like methyl-chlora-phenoxy-acetic acid, hexazinore, etc., and biological control by means of introducing *Pila globosa* (trophical snail), Chinese grass carp (fast growing fish) etc. that feed on many aquatic plants.

- Draining the water: Water present in the lake must be cleaned or drained completely.
- Desiltation: Dredging of the sediments in the lake to improve the soil permeability, water holding capacity and ground water recharge. Recent technological developments do permit wet dredging. Studies in Kolar district reveal that desilting of waterbodies helps in lowering fluorosis in borewell water (ground water).
- Constructed Engineered Wetlands: A constructed wetland is a water treatment facility that has gained importance in recent years for treatment of lakes. Duplicating the processes occurring in natural wetlands, constructed wetlands are complex, integrated systems in which water, plants, animals, microorganisms and the environment (sun, soil and air) interact to improve water quality. Constructed wetlands mimic nature by mechanically filtering, chemically transforming, and biologically consuming potential pollutants in the wastewater stream. These are shallow pools constructed on non-wetland sites as part of the stormwater collection and treatment system. They provide conditions suitable for the growth of emergent marsh plants. These systems are primarily designed for the purpose of stormwater management and maximum pollutant removal from surface water flows through physical, chemical and biological mechanisms. They are often used in sequence with a sediment basin or stormwater pond.

As an extension of the restoration programme, watershed management practices are essential for proper land use, protecting land against all forms of deterioration, conserving water for farm use, proper management of local water for drainage, flood protection and sediment reduction and increasing productivity from all land uses. Key steps for best management practices include:

- Pollution alleviation practices to reduce the engendering of non-point source of pollution (mainly agricultural and storm runoff) through source reduction, waste minimisation and process control.
- Afforestation with native species in desolate areas around the wetland (catchment area) to control the entry of silt from run off.
- The shorelines of the lakes are lined with bricks or stones in an attempt to control shoreline erosion.
- Constructed wetlands for the purpose of stormwater management and pollutant removal from the surface water flows.
- Infiltration trenches for reducing the storm water sediment loads to downstream areas by temporarily storing the runoff.
- Extended detention dry basins for removing pollutants primarily through the settling of suspended solids.
- Gyration of crops rather than monocultures to reduce the need for N and assist with pest control and help in aeration of soil.
- Promoting public education programs regarding proper use and disposal of agricultural hazardous waste materials and regular monitoring of lakes, which are rudimentary.

Watershed Ecology

An area of land that drains water, sediment and dissolved materials to a common receiving body or outlet. The term is not restricted to surface water runoff and includes interactions with subsurface water. Watersheds vary from the largest river basins to just acres or less in size. The study of watersheds as ecosystems, primarily the analysis of interacting biotic and abiotic components within a watershed's boundaries.

Ecosystem

A functioning natural unit with interacting biotic and abiotic components in a system whose boundaries are determined by the cycles and flux of energy, materials and organisms. It is valid to describe different ecosystems with different, overlapping sets of boundaries in the same geographic area (e.g. forest ecosystems, watershed ecosystems and wetland ecosystems). A watershed is just one of many types of ecosystems.

Watershed ecology is essential knowledge for watershed managers because it teaches us that watersheds have structural and functional characteristics that can influence how human and natural communities coexist within them. The gross structure of a watershed - its headwaters area, side slopes, valley floor, and water body, as well as its soils, minerals, native plants and animals - are, in one sense, raw material for all the human activities that may potentially occur there. The watershed's natural processes - rainfall runoff, groundwater recharge, sediment transport, plant succession, and many others - provide beneficial services when functioning properly, but may cause disasters when misunderstood and disrupted. It is crucial for people to understand watersheds and how they work before they make decisions or take actions that may affect important watershed structural or functional characteristics.

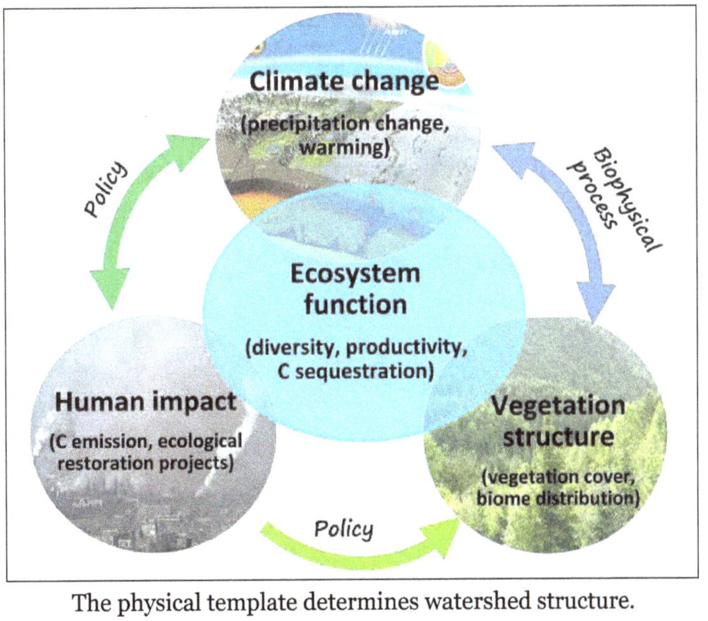

The physical template determines watershed structure.

Within the watershed, various forms of matter, including water, are in constant cyclic flow. Through these processes, an abiotic (non-living) template of air, water, and soil is formed, upon which life can exist. The physical template of watershed structure is ultimately determined by varying combinations of climatic, geomorphic, and hydrologic processes.

Climatology, the science of climate and its causes, becomes important in understanding regional issues in watershed science. Though sometimes used synonymously with weather, climate is actually a distinct term with important ecological ramifications. Climate refers to an aggregate of both average and extreme conditions of temperature, humidity, and precipitation (including type and amount), winds, and cloud cover, measured over an extended period of time. Weather refers to present-day environmental conditions; current temperatures and meteorological events make up weather, not climate. Long-term weather trends establish averages which become climatic regimes. Climate heavily influences watershed vegetation communities, streamflow magnitude and timing, water temperature, and many other key watershed characteristics.

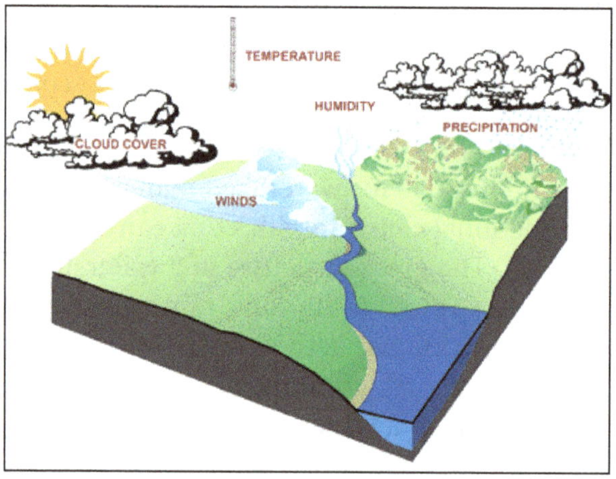

Climatic factors

Geology is defined as the science centered around the study of various earth structures, processes, compositions, characteristics, and histories. Geomorphology, however, refers specifically to the study of the landforms on the earth and the processes that change them over time. Fluvial geomorphology, referring to structure and dynamics of stream and river corridors, is especially important to understanding the formation and alteration of the stream or river channel as well as the flood plain and associated upland transitional zone; this is a critical discipline for effective, long-term watershed management.

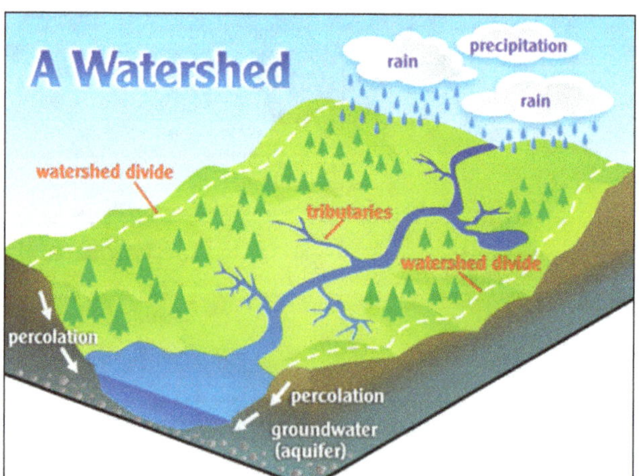

The physical template determines watershed structure. Geomorphology helps explain river and watershed form.

One of the life-sustaining cycles we are most familiar with is the hydrologic cycle. This cycle is a natural, solar-driven process of evaporation, condensation, precipitation, and runoff.

Hydrology is the science of water, as it relates to the hydrologic cycle. More specifically, it is the science of water in all its forms (liquid, gas, and solid) on, in and over the land areas of the earth, including its distribution, circulation and behavior, its chemical and physical properties, together with the reaction of the environment (including all living things) on water itself. The global water budget adds further insight into the water resources of our planet.

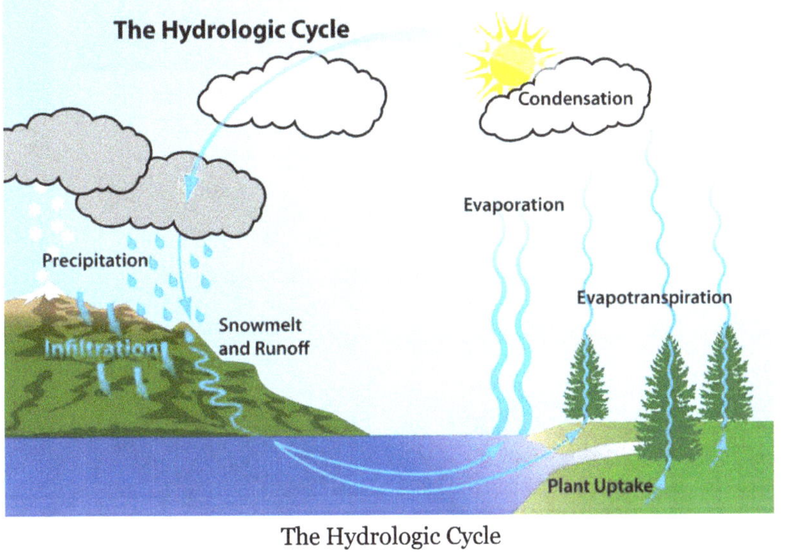

The Hydrologic Cycle

Table: The global water budget.

Saltwater	Percent of global water	Freshwater	Percent of global water
Oceans	97.542	Glaciers/Ice	1.806
Salt lakes/seas	0.004	Ground/Soil: root zone 0-2500 feet 2500-12500 feet	0.001 0.303 0.331
		Fresh Lakes Reservoirs	0.01 <0.001
		Atmosphere	0.001
		Rivers	<0.001
Total Salt	97.546%	Total Fresh	2.454%

The three elements of the physical template and other factors also interact significantly in determining the structure and composition of a watershed and its biotic communities. In this diagram, the strength of influence on the watershed is represented in two ways: the relative size of the arrows, and the level of control indicated at the right side of the diagram.

As a result of different combinations of these formative processes, different types of watersheds are created. Here are some examples that show how different from one another watersheds of different origin and physical template conditions can be.

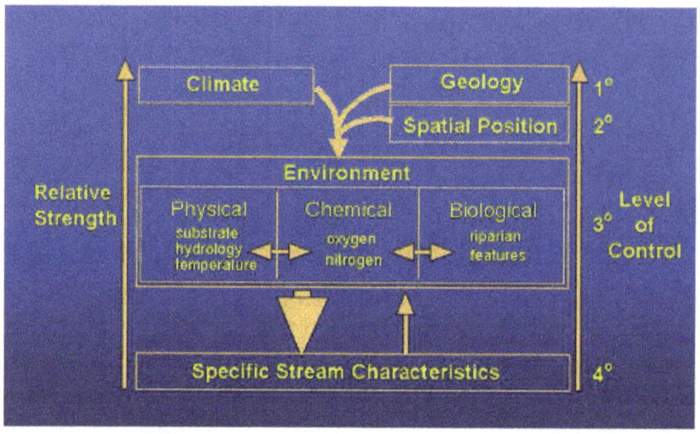

How elements of the physical template interact.

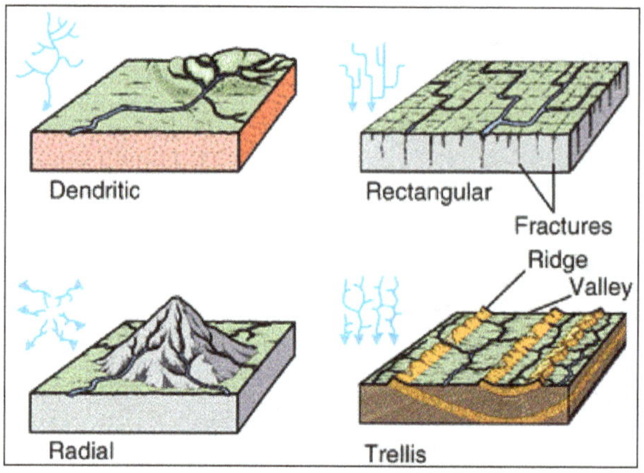

Valley Type II, moderately steep, gentle sloping side slopes often in colluvial valleys. Valley Type VI, moderately steep, fault controlled valleys. Valley Type VIII, wide, gentle valley slope with well developed floodplain adjacent to river terraces.

Biological Setting

Concepts of basic ecology provide us with the vocabulary to understand and describe the biological setting of watersheds and the interaction of biotic components with the physical template just described. It is important to have a working knowledge of the following basic ecological terms: ecology, species, population, community, habitat, niche, ecosystem, ecotone, biosphere. Beyond these basic definitions it is also important to understand the following ecological concepts: life history strategies, carrying capacity, competition, and symbiosis. You should become familiar with these basic definitions and key concepts before we discuss specific areas of ecology that are particularly relevant to watersheds.

Soil Ecology in Watersheds

Soil is a complex mixture of inorganic materials (sand, silt, and clay), decaying organic matter, water, air, and a great array of organisms. Because of its abundance of living organisms, soil is discussed here along with other "biological setting" components, even though soil is sometimes incorrectly described as a physical, non-living entity.

Soil has three basic properties which aid in its identification and taxonomy: color, structure, and texture. Soils often vary substantially from place to place within a watershed, and among different watersheds. To describe their differences, soils are classified into soil orders. Knowing the basic differences among types of soils can be useful for understanding why they vary in their suitability for supporting different land uses and ecological communities. Histosols, for instance, are poorly drained with a high organic content. These, essentially, are wetland soils - soils where a constant presence of water allows for the growth of water-tolerant plants, but breakdown of organic matter is slow due to the anaerobic condition of the soil. Oxisols, on the other hand, occur throughout the tropics - these soils are low in nutrients because most available nutrients are being utilized. The extremely high turnover rate (of matter and particularly nutrients) in the tropics makes these soils, when taken alone, highly infertile. This is one of the key reasons why deforestation in tropical watersheds creates a serious dilemma: once the vegetative cover is removed, there is little capacity for the site to self-restore.

Different soil horizons vary in appearance.

Food Webs and Trophic Ecology of Watersheds

Terrestrial and aquatic ecosystems have characteristic trophic (feeding) patterns that organize the flow of energy into, through and out of the watershed ecosystem and support the growth of organisms within the system. Food "chains" are rarely linear, hence the term food web, often used to describe the trophic interactions of organisms within an ecosystem.

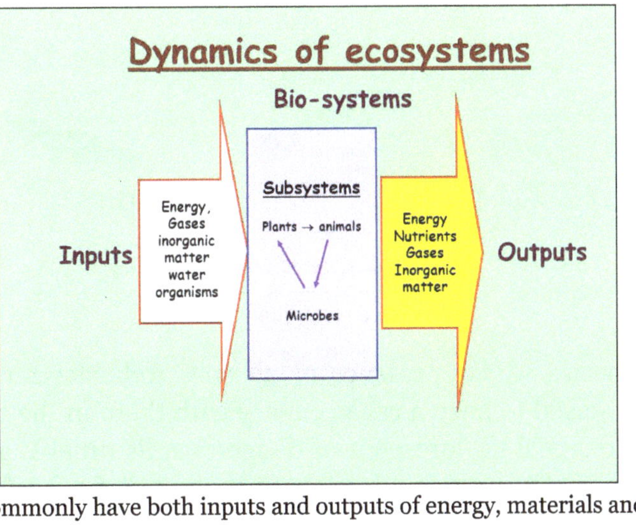

Ecosystems commonly have both inputs and outputs of energy, materials and organisms.

Within a food web, organisms interact and, in the process, may directly or indirectly affect other organisms. The example pictured is a simple, aquatic-only food web; when the whole watershed's terrestrial components are also considered, food webs can be very complex with numerous interactions among land-based and water-based species. Food webs also often recognize the different roles species play by terming them producers (organisms that generate food, primarily through photosynthesis), consumers (first-order consumers are vegetarians, second-order consumers feed on first-order, etc.), and decomposers (which feed on dead tissue and return nutrients and energy to other parts of the cycle), among other terms.

Species with especially far-reaching effects on an ecosystem are called keystone species. These species differ from dominant (i.e. abundant) species in that their effects are much larger than would be predicted from their abundance. They have a disproportionate effect on the composition of communities and ecosystem function. A keystone species' presence is often the lone reason for the presence of other organisms and/or the maintenance of unique ecological areas. The effects of keystone species are context-dependent, meaning that a species is not always a dominant controlling agent across its entire range, through all stages of its life cycle, or at all times of the year. The American Alligator is an example: it is not the most common species in southern swamplands and bayous, but it is an important predator that also modifies aquatic habitat structure by creating 'gator wallows'.

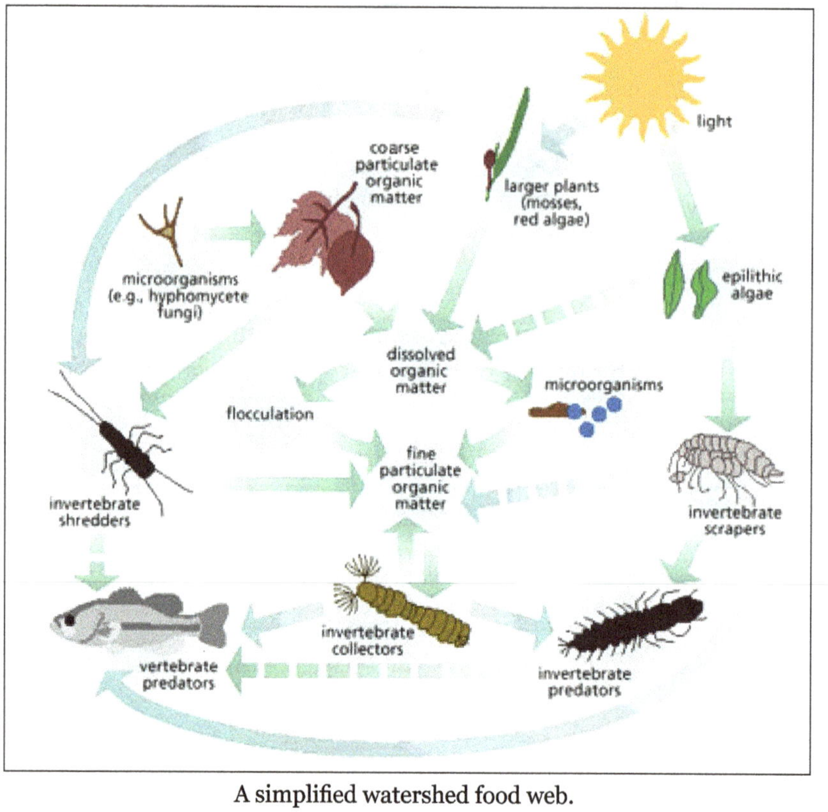

A simplified watershed food web.

Indicator species are species whose presence or absence indicates an environmental change. 19th-Century coal miners used to keep a caged canary with them in the mine shaft; the especially-sensitive canary would signal the presence of dangerous, flammable gases by ceasing to sing, and dying. The ever-cognizant miners could evacuate upon noticing the death of the pet bird. This is the concept behind an indicator species: they are modern-day canaries-in-a-coal mine. Watershed "canaries in a coal mine" include several types of aquatic invertebrates that are labeled "intolerant" of poor water quality, and amphibians such as frogs and salamanders.

Examples:

Amphibians: Amphibian declines, worldwide, are examples of indicators of some man-induced global changes. Various characteristics of amphibians make them particularly susceptible to environmental change (e.g. life both in water and on land, porous skin, a notably omnivorous diet). Localized extinctions, extirpations, or declines have been studied and have often shown a response to an anthropogenic disturbance. Cascade frogs, for instance, have been declining throughout the Cascade Mountains of Oregon. Biologists have attributed this species' demise to increased levels of ultraviolet light due to ozone depletion. These frogs lay their eggs in a clear, aqueous gel just under the surface film of alpine lakes, making them susceptible to the effects of the increased UV-B.

Benthic macroinvertebrates: Stream and lake invertebrates exhibit varying tolerances to different environmental factors. Some are more sensitive than others to sedimentation, temperature changes, and food availability. Many taxa are categorized as "intolerant" of many common pollutants. As such, the presence (or absence) of certain aquatic macroinvertebrates may serve as a sentinel for watershed condition. The Index of Biotic Integrity (IBI) was created as a method for assessing stream condition. The IBI creates a scale based on reference condition, centering around ratios of certain expected groups of larval insects of varying pollution tolerance.

Biodiversity (Genetic, Population, Species, Habitat)

Biodiversity is a contemporary term which has several subcategories. In general, the term applies to the relative amount of biological elements existing within a given area. Figure illustrates the complexity and importance of biodiversity at different levels from the genetic to the landscape scale, noting how biodiversity influences and helps define the structural, functional and compositional nature of our environment.

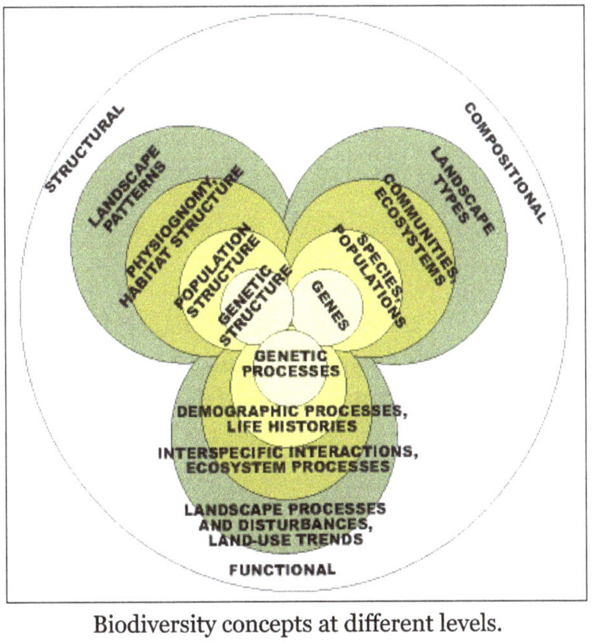

Biodiversity concepts at different levels.

Genetic biodiversity refers to the total number of genotypes available within a given population. For example, whooping cranes were driven to the brink of extinction; at one point the total global population stood at 14 individuals. Today, the population has returned to a more comfortable level. Still, the current population is limited to the genetic material which was contained within those 14 birds, and it will take eons and many, many generations for genetic diversity to build up again.

Populations with low genetic biodiversity may be more susceptible to certain diseases given the limited amount of genetic resistance potentially available. A "genetic bottleneck" refers to the loss of valuable survival traits from a population that has shrunk to a low level and then re-expanded.

Population biodiversity refers to the total amount of populations a given species has, worldwide. For instance, Pacific salmon are anadromous, meaning that they are born in freshwater, spend their adult life in the ocean, and then return to the fresh water from whence they originally came to reproduce and then expire. These fish rarely stray to other river systems, migrating in distinct populations from river to sea and then back to the same river.

While the total number of pink salmon may be "healthy", given the number of fish surviving in Alaska, population biodiversity may suffer if several rivers in southern British Columbia suddenly experience the loss of the runs of these fish.

Species biodiversity is the total number of species found within a given area. In natural systems, as an example, species biodiversity is considered quite high in the tropical regions of the world, while the number may be quite moderate in temperate zones. In the woodlands of Pennsylvania, for instance, it is not uncommon to count on one hand the total number of tree species within an acre of land. In the tropics, the number of tree species found within an acre of land may be over 250.

Habitat or ecological biodiversity refers to the number of different habitats or ecotypes found within a given region. In the Pacific Northwest, industrial forestry has reduced entire landscapes to monocultural tree plantations with a simple, homogeneous forest structure. On the other hand, in the region's natural systems there exists a much higher level of ecological diversity, given the various natural processes (e.g., wind, fire, flooding, disease, succession, competition) which create a mosaic of habitat types.

Natural Systems

The interactions and natural processes that link these abiotic and biotic components of watersheds exhibit what can be called system-like behavior.

The dictionary defines a system as "a group of interrelated, interacting, or interdependent constituents forming a complex whole." We have seen that natural systems such as watersheds have interacting components that together perform work (e.g. transport sediment, water, and energy) and generate products (e.g. form new physical structures like floodplains or channels, and form biological communities and new energy outputs). In a natural system, interactions make the whole greater than the sum of its parts - each of the physical and biological components of watersheds if they existed separately would not be capable of generating the work and the products that the intact watershed system can generate.

The natural systems concept is key to watershed management because it emphasizes that a watershed, as a natural system, is more than just a variety of natural resources coincidentally occurring in one place. Severely degraded watersheds may have lost several of their components and functions and provide fewer benefits to human and natural communities as a result. Thus it is clear that recognizing the natural system and working toward protecting the system's critical components and functions are key to sustainable watershed management. Other ecological concepts and theories help explain the idea of natural systems. These include spatial and temporal scale, disturbance theory, and the river continuum concept, all discussed below.

Spatial and Temporal Scale

Spatial and temporal scale are important parts of the natural systems concept. The spatial scale at which a system operates is an important factor in recognizing the system's key components, and in turn, the kind of management practices that may be appropriate. In figure, note the general tendency for systems at a larger spatial scale to have natural processes that operate on longer time frames.

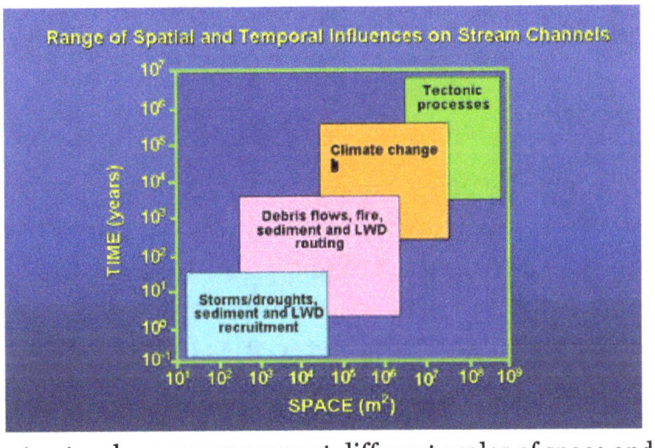

Different natural processes occur at different scales of space and time.

Disturbance Theory

Whereas natural systems have a certain degree of organization and order, they also exhibit constant change and disturbance at varying levels. Disturbance ecology often centers around a concept known as the intermediate disturbance hypothesis. This hypothesis explains why diversity is often highest in systems with intermediate levels of disturbance. Few species are capable of colonizing an area that either experiences high frequency or intensity of disturbance (e.g. frequent or intense flooding). In areas of low or infrequent disturbance, a small number of species optimally suited to local conditions establish themselves and outcompete other potential colonizers, so here too diversity tends to be lower.

The importance of natural disturbances in shaping landscapes and influencing ecosystems is well-documented in the scientific literature. Ecologists generally distinguish between relatively small, frequent disturbances and large, infrequent, so-called "catastrophic" disturbances. Much has been recently learned of the former, while a relative paucity of data exists on the latter.

Examples of small, frequent disturbances include seasonal floods, periodic grassland fires, and mild to moderate storms which periodically influence the landscape (e.g. wind-created forest canopy gaps). Examples of the large, infrequent disturbances include volcanic eruptions, hurricanes, and major wildfires. Recent examples of these include the 1988 Yellowstone fires (which charred over 1,500 mi2 of predominantly forested National Park land), Hurricanes Hugo and Mitch (which flattened trees and created landslides in Puerto Rico and Honduras, respectively), and Mount St. Helens, which erupted in May 1980 and affected more than 250 mi2 within the affected landscape, including areas of pyroclastic flow, debris avalanche, mudflow, blowdown, singe, and ash deposition.

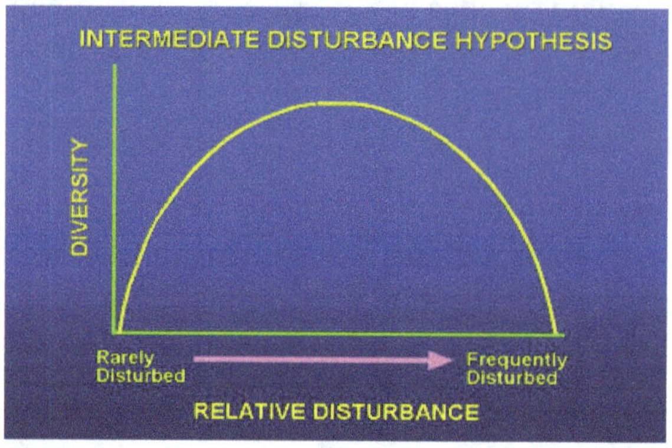

Intermediate levels of disturbance tend to favor biodiversity over too much or too little disturbance.

River Continuum Concept

This concept is a generalization of the physical and biological patterns often seen in different zones of rivers from source to mouth. Conceptually, from headwaters to outlet, there exists in a river a gradient of physical conditions width, depth, velocity, flow volume, temperature, and other factors. Geomorphologists have shown that lotic (flowing water) systems show patterns, or adjustments, in the relationship of a number of physical characteristics (e.g. stream width, depth, velocity, bed-load) along their entire length. Biotic characteristics in each zone reflect the influence of the physical influences they exist under; in other words, similar natural systems often develop under similar conditions. And as we move from the headwaters to a downstream reach, we see a continuum of physical conditions and a subsequent response in expected biota within these aquatic systems.

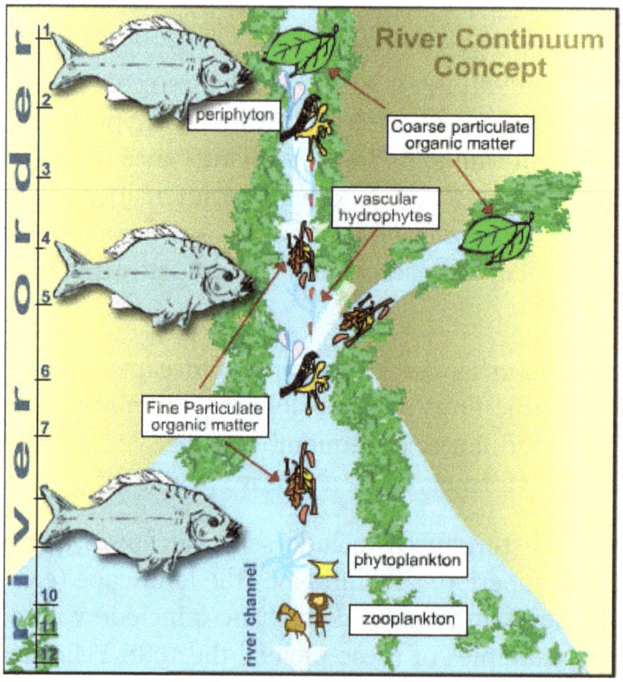

The river continuum concept.

Watershed Structure

The physical and biological components of watersheds and considered that together they comprise organized, functional systems. Basically, this includes structure of flowing waters (mainly rivers and streams with associated riverine wetlands and riparian zones), still waters (lakes and associated basin-type wetlands and shorelands), and upland areas of watersheds.

Flowing (Lotic) Systems

The US has more than 3.5 million miles of flowing water systems, which include springs and seeps, rivers, streams, creeks, brooks and side channels.

The Four-Dimensional Concept recognizes that lotic systems' structure exists in a four-dimensional framework, as below:

Longitudinal (in an upstream and downstream direction) - Flowing water systems commonly go through structural changes en route from their source to mouth. Three zones are usually recognized headwaters, where flow is usually lowest of anywhere along the system, slope is often steepest, and erosion is greater than sediment deposition; transfer zone, the middle range of the stream where slope usually flattens somewhat, more flow appears, and deposition and erosion are both significant processes; and the downstream end's depositional zone, where flow is highest but slope is minimal and deposition of sediment significantly exceeds erosion most of the time.

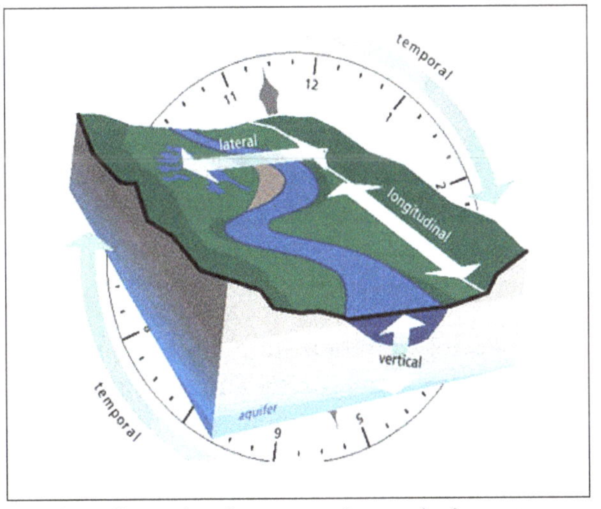

A four-dimensional concept of watershed structure.

Lateral (across the channel, floodplains and hillslopes): Again, significant variation occurs among stream types, but a common pattern includes the channel, the deepest part of which is called the thalweg; low floodplains that are flooded frequently, and higher floodplains (e.g., the 100-year or 500-year) that are rarely inundated; terraces, which are former floodplains that a downcutting stream no longer floods; and hillslopes or other upland areas extending up-gradient to the watershed boundary.

Vertical (surface waters, ground water and their interactions): It is always important to recognize that water bodies are not purely surface features; rivers and streams constantly interact with groundwater aquifers and exchange water, chemicals, and even organisms. Over its entire length, a stream often varies between influent reaches where surface water leaks downward into the aquifer, and effluent reaches where the stream receives additional water from the aquifer.

Temporal (through time, from temporary response to evolutionary change): The dimension of time is important because rivers and streams are perpetually changing. Structure as described in the other three dimensions above should never be considered permanent, and watershed managers should always think of structure not just as what is there now, but in terms of the structural changes in progress and their rates of occurrence.

Structure in Upland Areas of Watersheds

The physical form of the uplands in watersheds can vary greatly, in ways beyond the scope of this discussion. Here we focus only on the distribution of and variations in vegetation and land use, which together create the element of watershed structure called landscape pattern. Vegetation and land use patterns in watersheds are known to have many significant influences on the condition of the water bodies they drain into.

Aerial views reveal the complexity of landscape pattern, which implies many traits of a watershed's ecological functions.

Landscape patterns: Landscape ecology offers a simple set of concepts and terms for identifying basic landscape patterns: matrix, patch, and mosaic. The ecological term matrix refers to the dominant (> 60 percent) land cover, while a patch is a non- linear area that is less abundant and different from the matrix. A mosaic is a collection of different patches comprising an area where there is no dominant matrix. Various patch types have been described in the Examples of Different Patch Types box below. Basically, the most obvious landscape patterns are formed by combinations of native vegetation communities, unvegetated areas, and land use patterns.

Landscape pattern change. The individual patches in a landscape can change, and so can the entire landscape change in pattern and/or composition. Disturbances and various landscape processes maintain a constant dynamic, referred to as a shifting mosaic. Some landscapes remain in a "dynamic equilibrium" and, although changing steadily from place to place, retain an important quality called mosaic stability. A well-managed forestry operation, for example, would exhibit over the long term a constantly shifting set of locations where mature forest occurred, but at the same time sustains the relative proportions of forested and nonforested land in the area. Or, a landscape may evolve toward a new type of pattern and composition (e.g., via timber clearcutting, suburban sprawl, abandonment and succession of agricultural lands back to forest, or landscape change due to disease, fire, or global warming). It is always important, when analyzing landscape pattern and landscape change, to remain aw are that the spatial resolution of your information (how small a landscape feature you can detect) may or may not be sufficient to detect all the landscape changes of possible significance that may be occurring.

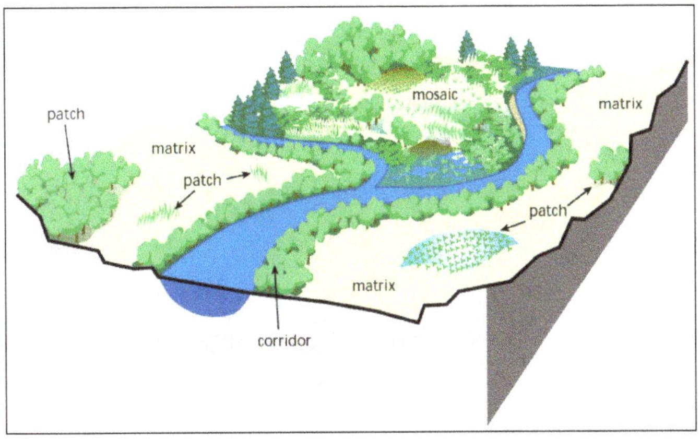

Useful units for describing landscape structure and pattern.

Vegetational patterns: Upland vegetation structure varies spatially, following various biogeographical patterns based on climate, physiography, soils, disturbance regimes, and their interactions. Vegetation communities are areas where a few species of plants dominate and establish a characteristic form or structure, within which a potentially large number of less abundant organisms also exist. Nationwide, there are hundreds of vegetation community types; the Society of American Foresters recognizes over 80 forest types alone. As a first step in analyzing vegetational patterns, it is easier to recognize a few generalized upland vegetation types based on their growth form, including:

- Forests (deciduous, evergreen and mixed);
- Shrublands;
- Grasslands;
- Forbs (broad-leaved herbs).

Examples of Different Patch Types:

- Disturbance patch: A patch which results from an alteration or disturbance.
- Remnant patch: A patch which survives disturbance for some extended period of time.
- Environmental patch: A patch which exists due to the natural variation of the environment (such as a soil type). Regenerated patch. This type of patch resembles a remnant patch, except that it has regrown on a previously disturbed site.
- Introduced patch: This type of patch is created by a human process (e.g. a clearcut, or a stand of planted trees).
- Corridor: A uniquely important type of patch that links other patches in the matrix (e.g. a river channel, riparian or buffer zone).

These categories are commonly found on land cover maps likely to be available in the GIS data for most watersheds, and can be consulted to give a general sense of vegetation patterns in the watershed.

Human activity has carved up and fragmented many of the natural vegetation patterns that formerly covered our watersheds. Without human influence, however, vegetation patterns would not be uniform due to different vegetation communities arising from different environmental conditions (e.g. variations in moisture and temperature due to slope and aspect) and events (e.g. fire, pest outbreak). In the West, the "rain shadow" is a common, basic example of how vegetation varies with physical position.

Examples of matrix, patch, and corridor.

As moisture-laden air is pushed east, it begins to slowly rise as it reaches the gradually- sloping foothills of the southern Sierra. As the air rises, pressure increases, and moisture in the air begins to condense. As the air continues eastward, it continues upward to the crest of the range, and moisture within the air is seemingly wrung-out, much like a sponge. The eastern slope of the Sierra is quite dramatic, topographically. Indeed, Mt. Whitney, the highest point in the coterminous United States, is just a matter of miles from North America's lowest point (Death Valley, CA). Much of the eastern slope is quite dry, since the air is now deprived of much of its moisture.

Land-use Patterns

Increasingly, the landscape structure and pattern we see is the result of widespread human activity. In all fields of environmental management including watershed management, analysis of land use types, patterns, and trends is commonplace. Because multiple uses occur in many locations and some land uses are not in themselves a visible landscape feature, mappers often use term land cover to describe the delineation of landscape structure and pattern formed by the dominant land uses and remaining vegetation communities. Some common land cover categories (indicating land uses within the areas) include:

- Urban land (residential, commercial, industrial, mixed).
- Agriculture (row crops, field crops, pasture).
- Transportation (roads, railroads, airports).
- Rangelands.

- Silviculture.
- Mining/extractive areas.

Like vegetation patterns, the land use patterns in a watershed can be studied through GIS data or maps. Human-dominated landscapes, just as natural landscapes, are shifting mosaics that often progress through a series of changes in what is dominant.

Pre-settlement landscapes in upstate New York were probably almost entirely forested. By the mid to late 19th century, clear-cutting had removed the forest from most areas. In the first part of the 20th century, cropland and pasture became widespread and dominated 70% or more of the landscape, with remnant scattered forest patches, wetlands and lakes. By the second half of this century, a trend in migration into cities and towns led to an assorted pattern of landscapes, including urban/suburban-dominated areas with forest and agricultural patches, heavily agricultural landscapes, and transitional rural lands with decreasing agriculture and increasing forest regrowth.

Watershed Functions

Sediment transport and deposition.

Transport and storage (of water, energy, organisms, sediments, and other materials): Because a watershed is an area that drains to a common body of water, one of its main functions is to temporarily store and transport water from the land surface to the water body and ultimately (for most watersheds) onward to the ocean. But, in addition to moving the water, watersheds and their water bodies also transport sediment and other materials (including pollutants), energy, and many types of organisms. It is important when recognizing the transport function to also recognize temporary retention or storage at different locations in the watershed.

Transport and Storage: As matter physically moves through the watershed, there are a number of terms which arise relative to various stages of cycling. Availability refers not just to the presence of an element in a system, but also speaks to the usability of a given agent. For instance, nitrogen gas may be plentiful in and around dam spillways, but N_2 is not a usable form for most aquatic organisms, and thus the availability of nitrogen is compromised. Detachment refers to the release of matter from an anchoring point, and its subsequent movement. Transport, a process most evident in stream channels, involves the movement of a material through a system. Deposition refers to a given endpoint within a cycle. Integration refers to the assimilation of matter into a site or organism following depositional processes. An example using these terms is included below.

Seston (suspended particulate matter) is an important food source for a few key aquatic organisms. In the context of the above continuum, seston's availability within a given stream system may refer to the source. In a hypothetical example, nitrogen and phosphorus, both limiting in many freshwater ecosystems, are made available to a stream during a period of elevated discharge. Attached algae (periphyton) begins to grow on rocks and submerged surfaces (availability). During growth and maturation, algal cells begin to slough off from the rocks and logs (detachment) and the current carries these cells, rich in chlorophyll, downstream (transport). Hydropsychid caddisfly larvae, which feed on seston, spin elaborate nets (much like a spider's web), anchoring these nets on cobble substrates or other submerged features. These nets catch the suspended particulates as they drift with the current (deposition). The caddisfly larvae then feed off particles captured in the net (assimilation); algal cells represent high quality food for these organisms and they may, as a population, exert an influence over the relative abundance of various particle-types.

Transport and Storage of Water

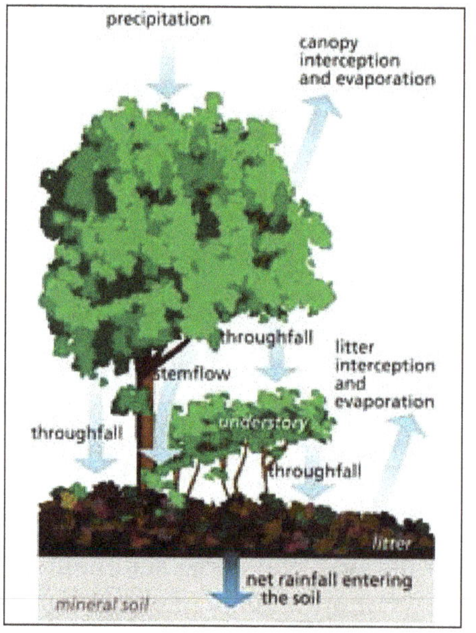

Dynamics of precipitation.

One can view a watershed as an enormous precipitation collecting and routing device, but transportation and storage of water actually involves a complicated mix of many smaller processes.

Even before precipitation reaches the ground, it interacts with vegetation. Trees and other vegetation are responsible for interception and detention of some of the rainfall, leading to some evaporation and also slowing the amount reaching the ground via through fall and giving it time for better infiltration to groundwater (one form of storage).

Saturation of soils, occurring when precipitation exceeds infiltration, leads to overland flow and, over longer time frames, drainage network development. The consistent flow of water in channels affects and shapes channel development and morphology in ways that seek dynamic equilibrium with the job to be done (moving water downstream).

Transport and Storage of Sediments

Watersheds also collect and transport sediments as a major function. Sediment transport and storage is a complex network of smaller watershed processes, like the water processes described above, and actually is inseparable from water transport and storage. Sediment related processes mostly involve erosion and deposition, but sediment transport and storage also play a longer-term role in soil development.

The drainage network development and channel development discussed above appears to be dominated by erosion at first glance, but the redeposition of sediments on floodplains is an important function that rejuvenates soils and influences the productivity and diversity of stream corridor ecosystems.

Cycling and Transformation

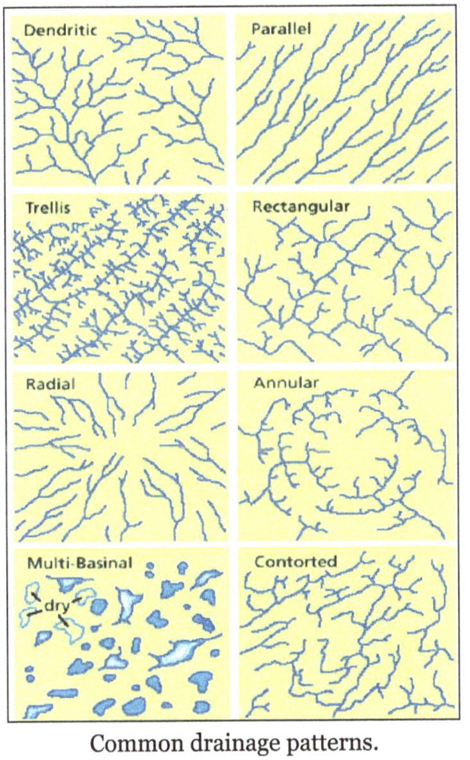

Common drainage patterns.

Cycling and transformation are another broad class of natural functions in watersheds. Various elements and materials (including water) are in constant cycle through watersheds, and their interactions drive countless other watershed functions. Figure, for example (overleaf), illustrates interactions of the carbon and nitrogen cycles with stream biota and the resulting influence on dissolved oxygen. Elements like carbon, nitrogen, and phosphorus comprise the watershed's most important biogeochemical cycles. Cycling involves an element of interest's transport and storage, change in form, chemical transformation and adsorption.

Nutrient Spiraling: The flow of energy and nutrients in ecosystems are cyclic, but open-ended. True systems, in both an environmental and energetic context, are either "open" (meaning that there is some external input and/or output to the cyclic loop) or "closed" (meaning that the system is self- contained). In watersheds, streams and rivers represent an open-system situation where energy and matter cycles, but due to the unidirectional flow, the matter does not return to the spot from whence it came. Also, nutrients "spiral" back and forth among the water column, the bodies of terrestrial and aquatic organisms, and the soil in the stream corridor en route downstream. Hence, the concept of nutrient "spiraling" implies both movement downstream and multiple exchanges between terrestrial and aquatic environment, as well as between biotic and abiotic components of the watershed.

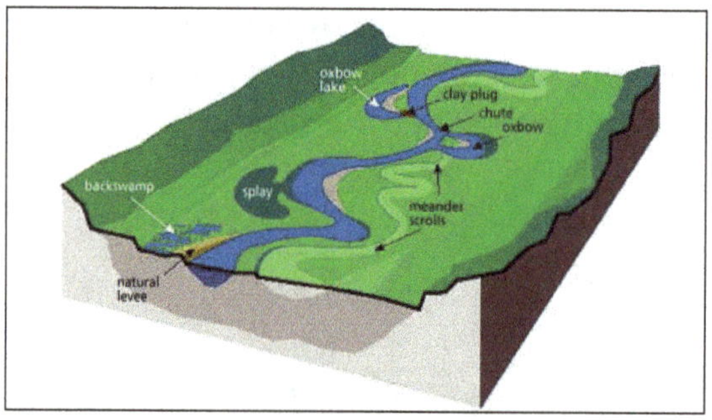

Common channel and near-channel features.

The Cycling of Carbon and Energy: In food webs, carbon and the subsequent synthesized energy is cycled through trophic (food web) levels. Energy transfer is considered inefficient, with less than 1 percent of the usable solar radiation reaching a green plant being typically synthesized by consumers, and a mere 10 percent of energy being typically converted from trophic level to trophic level by consumers.

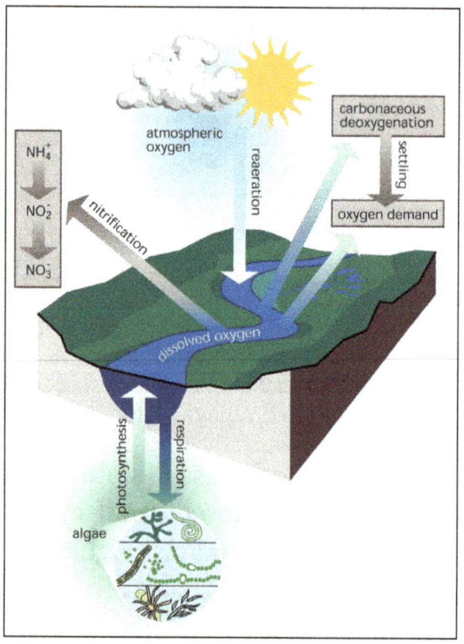

Carbon and nitrogen cycles affect stream biota.

Nitrogen (N): N_2 (gaseous state) is not usable by plants and most algae. N^-fixing bacteria or blue-green algae transform it into nitrite (NO_2) or ammonia (NH_4). N^- fixation, precipitation, surface water runoff, and groundwater are all sources of nitrogen. Under aerobic conditions, NH_4^+ is oxidized to NO_3^- (nitrate) in the nitrification process. Losses of N occur with stream outflow, denitrification of nitrate (NO_3) to N_2 by bacteria, and deposition in sediments. Unlike P, inorganic N ions are highly soluble in water and readily leach out of soils into streams. NH_4^+ (ammonium) is the primary end-product of decomposition.

Phosphorus (P): Phosphorus in unpolluted watersheds is imported through dust in precipitation, or via the weathering of rock. Phosphorus is normally present in watersheds in extremely small amounts; usually existing dissolved as inorganic orthophosphate, suspended as organic colloids, adsorbed onto particulate organic and inorganic sediment, or contained in organic water. Soluble reactive phosphorus (consisting of ionic orthophosphates) is the only significant form available to plants and algae and constitutes less than 5 percent of the total phosphorus in most natural waters. Phosphorus tends to exist in waters of a pH of 6-7. At a low pH (<6), P tends to combine readily with manganese, aluminum, and iron. At a higher pH (>7), P becomes associated with calcium as apatite and phosphate minerals. It is normally retained in aquatic systems by algae, bacteria and fungi.

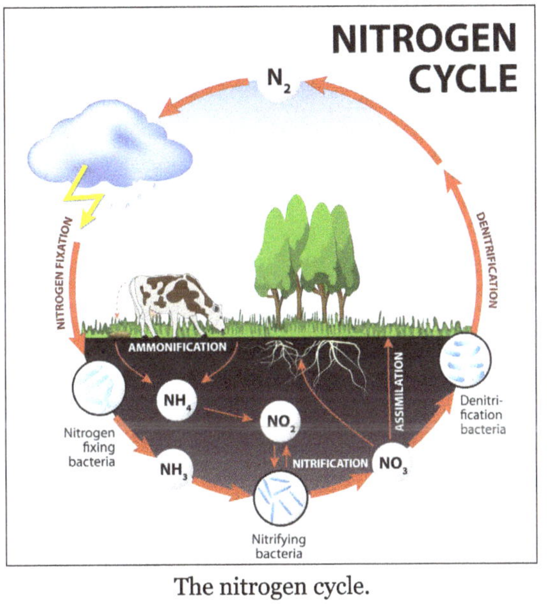

The nitrogen cycle.

Nitrogen and Phosphorus limitation: Most watershed systems (both the aquatic and terrestrial realms) are either N or P limited, in that these are the required elements which are at the lowest availability. As a general rule, the N:P ratio should be 15:1. A lower ratio would indicate that N is limiting, a higher ratio places P in that role. Commonly P is the limiting factor. Often, the slightest increase in P can trigger growth, as in algal blooms in an aquatic setting. In N and P limited systems, an input of either element above and beyond normal, "natural" levels may lead to eutrophication.

The stream corridor is often a mediator of upland-terrestrial nutrient exchanges. As N and P move down through subsurface flow, riparian root systems often filter and utilize N and P, leaving less to reach the stream. This has a positive influence on those already nutrient- overloaded bodies of water, but would not necessarily be a positive influence on organisms struggling to find food in very clean, nutrient-limited headwaters streams. Microbes also denitrify significant amounts of N to the atmosphere. Still, N-fixers, like alder, may serve as sources of N for the stream channel, and groundwater pathways between the stream and the streamside forest may provide significant quantities of nitrogen.

Decomposition

Decomposition involves the reduction of energy-rich organic matter (detritus), mostly by microorganisms (fungi, bacteria, and protozoa) to CO_2, H_2O and inorganic nutrients. Through this process they both release nutrients available for other organisms and transform organic material into energy usable by other organisms. In lakes, much of the decomposition occurs in the waters prior to sedimentation. In the headwater reaches of streams, external sources of carbon from upland forests are a particularly important source of organic material for organisms and decomposition of microscopic particles occurs very rapidly. The bacteria and fungi modify the organic material through decomposition and make it an important food source for invertebrate and vertebrate detritivores, thereby reinserting these nutrients and materials into the watershed's aquatic and terrestrial food webs.

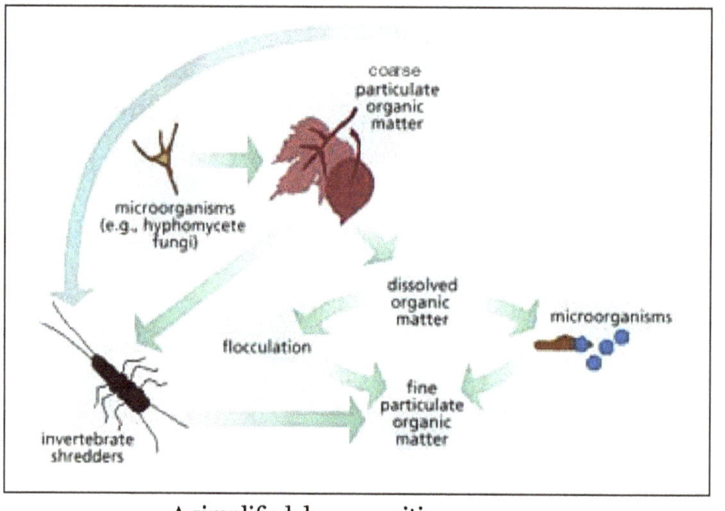

A simplified decomposition process.

Decomposition is influenced by moisture, temperature, exposure, type of microbial substrate, vegetation, etc. Specifically, temperature and moisture affect the metabolic activity on the decomposing substrate. Nutritional value (as well as palatability) of the decomposing structure will also affect the time involved in complete breakdown and mineralization. Decomposition involves the following processes:

- The leaching of soluble compounds from dead organic matter.
- Fragmentation.
- Bacterial and fungal breakdown.
- Consumption of bacterial and fungal organisms by animals.
- Excretion of organic and inorganic compounds by animals.
- Clustering of colloidal organic matter into larger particles.

The process of death and consumption, along with the leaching of soluble nutrients from the decomposing substrate, release minerals contained in the microbial and detrital biomass. This process is known as mineralization.

Ecological Succession

The classical ecological definition of plant succession involves a predictable set of vegetative changes through a series of discrete stages (seres). Recent challenges to the original succession concept suggest that succession does not necessarily involve a "climax" stage (after which additional changes in dominant species and structure do not normally occur). Within the watershed, succession may vary with spatial scale, elevation, and topography. Modern successional theories view the landscape as being in a sort of dynamic equilibrium, in that various patches make up a shifting mosaic of various successional stages.

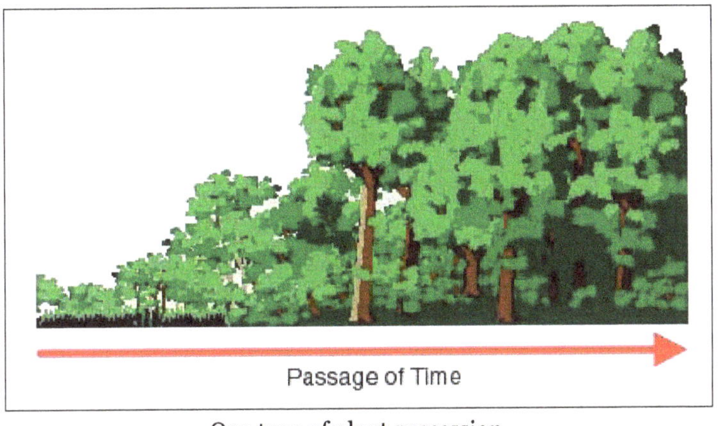

One type of plant succession.

In watershed terms, succession is a process that circulates significant amounts of the watershed's energy, water and materials from the abiotic environment back into the biotic, and from one set of predominant organisms on to a subsequent set of dominant organisms. Characteristic forms of succession may be typical of specific parts of the watershed. Succession builds and gradually changes vegetational structure that serves many critical functions such as maintaining varied habitat (recall the earlier discussion of the highest biodiversity often being found in areas of intermediate disturbance) and reestablishing renewable resources for human use, like woodlots.

Example: Terrestrial Forest Succession. Much of the western slope of the Sierra-Nevada Range is generally classified as a mixed- conifer forest. Timber species like ponderosa pine, white fir, and incense-cedar occur in varying proportions across the landscape continuum. Historically, wildland fires burned in relatively frequency, maintaining the forest in varying stages of growth, development, and species composition. The fire-resistant pines would dominate in areas of relatively frequent fire disturbances; fir and cedar would dominate in areas where fire was somehow excluded. High intensity disturbances (e.g., the so-called stand-replacing events), like large fires, might destroy the majority of trees, thereby reverting the landscape to a grassland or chaparral community. A simple view of Sierra-Nevada forest succession is as follows:

- Early successional stage: Following a catastrophic fire, Sierran wildlands are usually chaparral or grassland communities. This is also referred to in a forest development context as the "stand initiation stage."
- Mid-successional stage: The brush and grass sere is gradually overtaken by shade- intolerant pines, which prefer growth in areas of open sunlight. Also known as the "stem exclusion stage."

- Mid- to Late- successional stage: If fire is excluded, shade-tolerant fir and cedar begin to occupy the pine understory. This corresponds to the "understory reinitiation stage."
- Late-successional stage: Also known in forest development circles as the "old growth stage", this sere involves the dominance of fir and cedar, with the occasional old-growth pine.

The California gold discovery of 1848 created a massive immigration into the region, and many of the Sierran forests were felled for building construction, firewood, and mine timbers. As a result, many Sierran forestlands quickly became chaparral and grassland communities. These clearcuts, coupled with the exclusion of fire beginning with the establishment of the U.S. Forest Service, are the reasons much of the Sierra is currently a second-growth forest of shade-tolerant forests dominated by white fir. Today, management plans in the Sierra center around re-establishing forest structure based on natural succession.

Example: Riparian Succession. Along the stream, riparian vegetation is subjected to periodic spates of flooding, erosion and redeposition. As such, the life history patterns and physiological adaptations of different tree, shrub and herbaceous species will enable them to dominate in different stages of riparian succession. The following are some common plant strategies for survival in the riparian corridor:

- Invaders, for instance, are typically the first trees to establish themselves following a landscape disturbance such as a flood or landslide. Examples of invaders are alder and aspen.
- Resisters are those species which are adapted to resist the stress of their environment. In riparian corridors, these are species which can withstand the force of intense floods. The flexible, "whippy" stems of willow serve as an example of a resister-like adaptation.
- Endurers are species which endure floods by timing or mode of reproduction. Cottonwood, willow, dogwood, and alder may reproduce by sprouting from a severed and deposited branch. Another example of an endurer would be the coast redwood which has the ability to sprout adventitious roots to withstand the effects of periodic burial.
- Evaders reproductively "wait out" the process of disturbance by producing seeds which are capable of experiencing dormancy through the given disturbance. Riparian trees do not exhibit this strategy, but many riparian understory species do.
- Avoiders find a way to survive disturbances by simply not growing in regions of frequent perturbation. In general, conifers take root and exist in riparian systems only in areas where the intensity and frequency of disturbances are low.

What results is a mosaic of patch types on a small scale within the riparian zone. Protected areas mixed with areas of frequent erosion and deposition create a patchwork of various soil types and successional stages within the stream corridor. Woody debris, large enough and decay-resistant enough to withstand fluvial processes, can create "islands" of refuge where conifers may eventually be able to take root again near the stream channel. This "process loop" between woody debris deposition, alder establishment, and eventually leading to conifer establishment (and the prospect of recurring woody debris deposition) represents an example of a riparian successional process.

References

- Limnology-Horne-Goldman, aseriann: nd.edu, Retrieved 23 June, 2019
- Lake-Ecosystems: researchgate.net, Retrieved 12 April, 2019
- Formation-of-lakes-with-diagram, lakes, geography: yourarticlelibrary.com, Retrieved 21 March , 2019
- Limpage, monograph, energy: ernet.in, Retrieved 24 March , 2019
- Watershedecology, modules, watertrain: epa.gov, Retrieved 7 May, 2019

5

Wetlands Ecology

Wetlands are ecosystems which are flooded by water and which contain aquatic plants adapted to the unique hydric soil. The topics elaborated in this chapter will help in gaining a better perspective about the conservation of wetlands as well as the different plants, birds and other organisms found in wetlands.

Wetlands

Wetland is a complex ecosystem characterized by flooding or saturation of the soil, which creates low-oxygen environments that favour a specialized assemblage of plants, animals, and microbes, which exhibit adaptations designed to tolerate periods of sluggishly moving or standing water. Wetlands are usually classified according to soil and plant life as bogs, marshes, swamps, and other similar environments.

Geographic Distribution of Wetlands

Wetlands are found all over the world in every biome, or major life zone. Some wetlands, such as tidal marshes, fit the definition of a transitional zone because they occur where open water and land meet. Others, such as prairie potholes of central North America and Carolina bays (elliptical depressions) of the Atlantic Coastal Plain, are fed mostly by precipitation or groundwater and are not associated with a distinct body of water. Still others, such as the bog-and-fen mosaic of the taiga (boreal forest), are dominant features of the regional landscape.

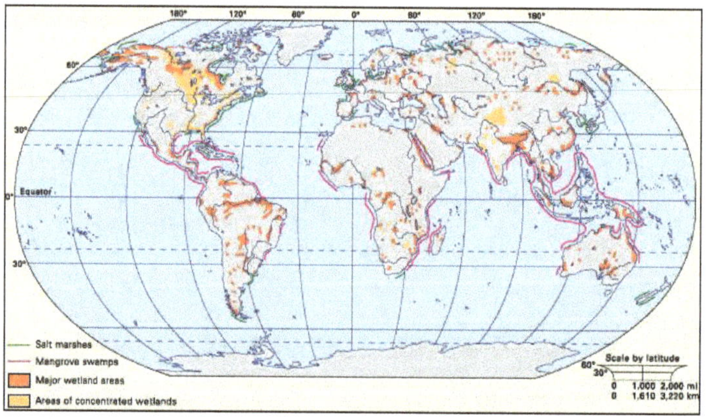

Wetland: Major wetland areas and worldwide distribution of salt marshes and mangrove swamps.

Wetlands are most abundant in boreal and tropical regions, though a wide variety of inland and coastal wetlands are also found in temperate regions. This distribution is generally due to conditions that promote an abundance of water. For example, the peatlands of Borneo and the Peruvian

Amazon occur within the tropical rainforest biome. In the treeless tundra of Alaska and Canada, saturated and flooded wetlands are underlain by permafrost. There, many parts of the landscape are made up of peatlands composed of black spruce (Picea mariana) and white spruce (P. glauca), which are sustained by rainfall or melting snow. Wetlands are also found in the hot desert biome—for example, the Mesopotamian marshlands found at the confluence of the Tigris and Euphrates rivers. In temperate regions, wetlands typically are found near coastlines, rivers, lakes, or other locations where local water input exceeds output.

Environmental Conditions

Climate

Wetland formation is influenced by climate patterns and the limitations posed by landforms. The net balance of precipitation and evaporation determines the quantity and timing of water available for the formation or maintenance of wetland conditions. Water flows downhill, and the geomorphology of the landscape determines where it gathers as well as what topography or subsurface layers prevent it from draining away. Each wetland has a water signature, or hydroperiod, which is characterized by the timing, duration, and quantity of water in the system. Furthermore, inundation has biological consequences, because it prevents atmospheric oxygen from being replenished in the soil. As a result, only organisms that can tolerate or are adapted to low-oxygen or anoxic (negligible oxygen) conditions have an advantage in wetland environments.

Soils

Peat bed

Wetland, or hydric, soils form when saturated or flooded conditions last long enough during the growing season to cause anaerobic (oxygen-depleted) regions to occur in the upper part of the soil, which includes the root zone. Such soils can be organic (containing organic compounds) or derived from minerals. Organic wetland soils, such as peatland soils, contain at least 12 percent organic matter and are typically acidic; they also possess a high water-holding capacity and low nutrient availability. Organic matter builds up in the soil when low oxygen conditions halt or slow decomposition. Mineral wetland soils, on the other hand, have less than 12 percent organic matter, and they often exhibit gleying, where ferric iron (Fe^{3+}) and manganese are reduced (that is, they gain electrons) in the soil by anaerobic bacteria thriving in the depleted oxygen conditions. The resulting ferrous iron (Fe^{2+}) becomes concentrated in a deep soil layer (soil horizon). In waterlogged soils, the topsoil and upper soil layers take on a black, gray, or blue-green colour. Pore linings (the coatings on the surface of open spaces in the soil) in wetland mineral soils are often red, because plant roots, which make many of the pores, release oxygen into the oxygen-depleted soil. The presence of this oxygen in an anaerobic environment oxidizes some of the ferrous iron remaining in the water and concentrates it along the pore linings. Plant litter and animal waste may occur at the surface of flooded mineral soils.

Wetland Functions and Ecosystem Benefits

Wetland functions are defined as the physical, chemical, and biological processes or attributes that are vital to the integrity of the wetland system. Because wetlands are often transition zones (ecotones) between terrestrial and deepwater aquatic systems, many processes have major implications for species. Since wetlands may provide food and habitat for many terrestrial and many aquatic species, wetland biodiversity is often higher than that of adjacent ecosystems. In addition, wetlands can affect the export of organic materials and serve as a sink for inorganic nutrients and atmospheric carbon. They play a major role in the biosphere by providing habitats for several plants, animals, and other forms of life; they may also serve as the last refuges for many rare and endangered species.

Some wetlands, such as swamps and marshes, are considered to be some of Earth's most productive ecosystems. To humans, wetlands are valuable for their sportfishing, hunting, and recreational uses. In addition, the capacity of wetlands to absorb a great amount of water also benefits developed areas, especially during periods of flooding. Wetland systems can also protect shorelines, recharge groundwater aquifers, and cleanse polluted waters. They have been described as "the kidneys of the landscape."

Abiotic Factors that Influence Wetland Ecosystems

Water

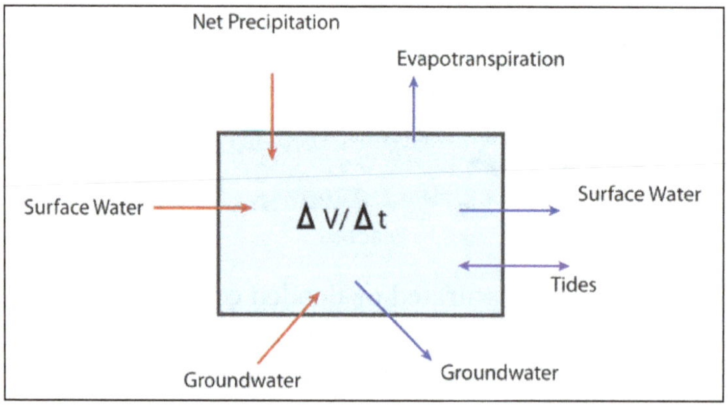

Hypothetical wetland water budget.

The movement, distribution, and quality of water is the primary factor influencing wetland structure and function. To be classified as a wetland, the presence of water must contribute to the formation of hydric soils, which are formed under flooded or saturated conditions persisting long enough for the development of anaerobic conditions during the growing season. Water conditions in wetlands can vary tremendously with respect to the timing and duration of surface water inundation as well as seasonal patterns of inundation.

In coastal wetlands, tidal influence drives the movement and distribution of water and can range from permanent flooding in subtidal wetlands to less frequent flooding in others, with changes in water level occurring daily or semi-daily. Inland wetlands, which lack daily tidal influences, can also be permanently flooded on one extreme or intermittently flooded on the other extreme, with fluctuations over time often occurring seasonally. It is the balance of water inflows and outflows, or the water budget, as well as the geomorphology and soils that determine the timing, duration, and patterns of flooding in a wetland.

A wetland's water budget describes its change in water volume over a given time interval, and includes all sources of water inflows (S_i = surface water, G_i groundwater, P_n = net precipitation, and T = tides) and outflows (S_o = surface water, G_o = groundwater, ET = evapotranspiration, and T = tides).

For most wetlands, the sources of inflows (e.g., precipitation, surface flow, groundwater flow, tides) and outflows (e.g., evapotranspiration, surface flow, groundwater flow, tides) change over time. As such, hydrology is rarely stable but fluctuates over time resulting in pulsing hydroperiods. Hydrologic pulses can alter productivity along a flooding gradient by altering the extent of flood subsidies and stresses in a wetland. When subsidies are high but stress is relatively low, pulses can promote productivity by introducing water, sediments, and nutrients while also removing waste materials and toxins.

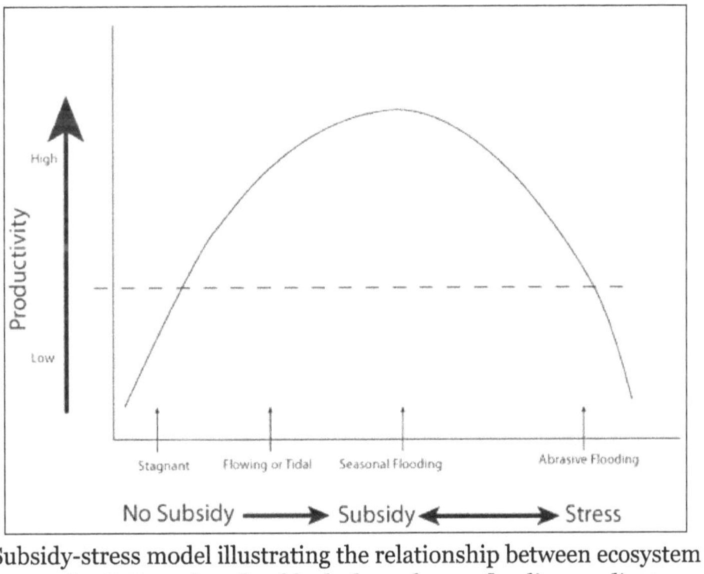

Subsidy-stress model illustrating the relationship between ecosystem productivity and wetland hydrology along a flooding gradient.

Flooding can affect the physiochemistry of wetlands in various ways. Water can introduce or remove sediment, salt, nutrients or other materials from wetlands, thereby influencing its soil and water chemistry. Hydrology also influences the structure and function of wetland ecosystems through its influence on species richness, productivity, rates of organic matter accumulation, and nutrient cycling. Hydrology may restrict species richness in areas subject to long-term flooding while enhancing it in areas with variable or pulsing hydroperiods. Similarly, productivity is typically lower in permanently flooded, stagnant wetlands, or in drained wetlands than in slow-flowing or seasonally flooded wetlands. The anaerobic conditions created under these inundated or flooded conditions often limit decomposition rates, thereby promoting organic matter accumulation in soils, and can alter reduction-oxidation reactions controlling nutrient transformations in wetland soils.

Productivity is low when flood pulses are minimal and water is stagnant, as well as when pulses are frequent and intense. When flood pulses are intermediate in frequency and intensity, productivity is maximized.

Oxygen Availability

The inundation or saturation of wetland soils by water leads to the formation of anaerobic conditions as oxygen is depleted faster than it can be replaced by diffusion. The rate of oxygen loss in flooded soils can vary depending on other soil conditions, such as temperature and rates of microbial respiration. In most wetlands, small, oxidized layers of soils may persist on the surface or around the roots of vascular plants, but generally, anaerobic, or reduced, conditions prevail.

The prevalence of anaerobic conditions in wetlands has a tremendous impact on their biogeochemistry, with important implications for carbon, nitrogen, phosphorus, iron, manganese, and sulfur transformations. Wetlands can function as sources, sinks, or transformers of these materials, depending on inflows, outflows, and internal cycling rates. One of the most important biogeochemical cycles in wetlands is the nitrogen cycle, and while the potential transformations are not unique to wetlands, the dominance of anaerobic transformations does set wetlands apart from other ecosystems. One such anaerobic transformation is denitrification, in which nitrate is lost to the atmosphere via conversion to nitrogen gas or nitrous oxide by bacteria. In many wetlands, nutrient availability is dramatically altered by agriculture or other practices that increase nutrient loading, contributing to changes in ecosystem structure and function. Through processes like denitrification and plant uptake, wetlands can help remove some of this excess nitrogen introduced to wetland and aquatic ecosystems.

Wetland Communities and Ecosystems

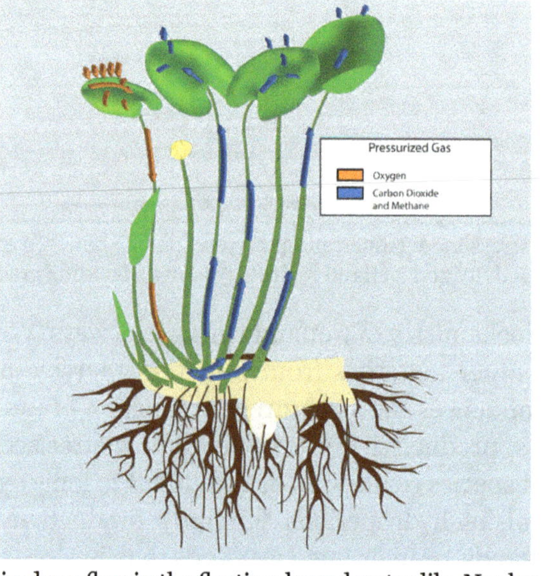

Pressurized gas flow in the floating-leaved water lily, Nuphar lutea.

Because of the predominance of water and anaerobic conditions in wetlands, the organisms living there, especially rooted plants, often exhibit remarkable adaptations to deal with the stresses imposed by flooding. These adaptations, including pressurized gas flow, creation of oxidized root zones, and anaerobic respiration, allow wetland plants to remain productive under otherwise stressful conditions, making wetlands among the most productive ecosystems in the world. This high primary production, in turn, supports high rates of secondary production, rates that can exceed those of terrestrial ecosystems.

Pressurized gas flow is one mechanism for overcoming oxygen root deficiency in plants growing in anaerobic wetland soils. Gaseous transport results in the flow of oxygen from the atmosphere to the roots, and carbon dioxide and methane from the roots to the atmosphere.

Wetlands exist along soil-moisture gradients, with wetter soils at lower elevations and drier soils at higher elevations. Wetland plant communities develop in response to this environmental gradient based primarily on their individual abilities to tolerate flooding and anaerobic soils but also in response to biotic interactions with other species. Establishment of plant species along an environmental gradient can contribute to sharp plant zonation patterns, as can be seen in coastal wetlands where species separate out along an elevation gradient in response to differences in flooding and salinity.

Typical plant zonation pattern in coastal marshes of the eastern Gulf of Mexico, Spartina alterniflora to Juncus roemerianus.

Salt marsh plant communities shift in dominance from the first to the second along an elevation gradient before transitioning into maritime pine uplands in Grand Bay National Estuarine Reserve, Mississippi, USA.

The development of these productive and often diverse plant communities fuels complex food webs that not only sustain microbial communities through large inputs of detritus to wetland soils but also support diverse communities of animals that utilize wetlands for part or all of their lives. Detritivores, such as shredding insects and crayfish, can utilize dead plant material as their primary energy source, while others (e.g., marsh periwinkle snails) help process organic matter for subsequent use by other organisms. Herbivory of algae by invertebrates and small fish and of plant biomass by some invertebrates, birds, and mammals (e.g., grasshoppers, geese and muskrats) is a significant energy source for primary consumers in many wetlands. Secondary production by these primary consumers supports higher trophic levels, including predatory insects, fishes, reptiles, amphibians, birds, and mammals.

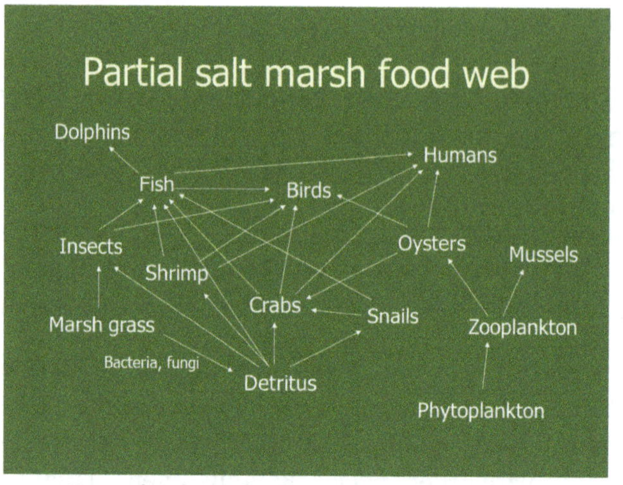

Example of a food web in a coastal salt marsh.

This food web of a marsh in Georgia, USA, lists the important primary producers, herbivores, and carnivores in order of importance.

As awareness of wetland ecosystem services and values has increased, wetland ecological research also has increased. Wetland ecologists examine interactions between species and their environment, recognizing the important role that hydrology plays in shaping the physicochemical environment and biological communities in wetlands. Within that framework, ecologists can examine a near-endless array of ecological topics, from the physiology of species coping with flooding stress and anoxia to species interactions, to the impacts of and feedbacks to global climate change. The diversity of wetland types, the biodiversity they support, and the numerous functions they provide make wetlands an exciting and rewarding arena in which to explore fundamental ecological questions. And it is this science that informs efforts to manage, restore, and conserve the wetlands of the world.

Organisms in Wetlands

Plants of the Wetlands

The Water Lily (Nymphaea violacea) with its blue, white, violet or pink petals and yellow stamens is one of the most common of the water lilies in the Top End. The large round two-lobed leaves float on the water and provide shelter and food for large numbers of insects and their larvae. The underside of the leaves and the stems are coated with a furry growth consisting of a tangled mixture of many different types of algae which are important as primary producers for small invertebrates.

The Water Snowflake (Nymphoides indica) and the Lotus Lily (Nelumbo nucifera) are two other common plants of the lagoons. The leaves of the Lotus Lily may be either emergent (held above the water) or floating. The emergent leaves follow the path of the sun during the day. Many of the truly aquatic plants have both submerged and floating leaves. Floating Pondweed is an example of such a plant. The stems of most water plants are not there to support the rest of the plant as for land plants, but simply to act as an anchor chain to the bottom.

Submerged plants may have long strap-like leaves (e.g. Ribbonweed (Vallisneria spp.)), leaves with undulating margins (e.g. Curly Pondweed) or finely divided leaves (Diocaulon sp. and Yellow Bladderwort (Utricularia gibba)). The Bladderwort is named after the little hollow globes or bladders growing on the leaves which trap aquatic invertebrates (microcrustaceans such as Daphnia) and so gain nutrients for the plant.

There are several types of small free floating plants. The most common is the fern-like Azolla spp. which can cover areas of still water like a green lawn, much to the dismay of anyone who steps onto the 'lawn'.

Grasslands

The grasslands are dominated either by annual species, such as Wild Rice (Oryza rufipogon) or perennials, such as (Hymenachne acutigluma). Both are very important to the ecology of the black soil floodplains. The rice is a major food source for Magpie Geese (Anseranas semipalmata) just as the young are hatching, and the Hymenachne forms dense floating mats of grass upon which Saltwater Crocodiles (Crocodylus porosus) can build their nests. Its seeds are also being used to rehabilitate tracts of bare land after Mimosa pigra has been removed by aerial spraying.

Other grasses of the wetlands are various Echinochloa species, one of which, Echinochloa preastans, joins with Hymenachne to make floating mats, and two species of Paspalum. The grasslands are located on the featureless black soil plains adjacent to the middle and lower reaches of the main rivers. Here the floodwaters are shallowest, and except for the occasional clumps of Pandanus (Pandanus spiralus), there is no other vegetation to be seen. The seeds of wetland grasses provide some species of birds with food. Large flocks of Little Corellas (Cacatua pastinator) may be seen feeding in the dry season in places where there seems to be nothing but an expanse of dead, dried out grass. Emerald Ground Doves (Chalcophaps indica) are also seen foraging for seeds on burnt ground.

Sedgelands

The Spikerushes (Eleocharis spp.) are most common of the sedges in the Top End wetlands. They grow in all other states of Australia and provide homes for a number of bird, snake and mammal species. Spikerushes grow from robust rhizomes that are a food source for a number of animals including Magpie Geese (Anseranas semipalmata) and it grows in waters up to 2 m deep. Rice Sedge is another common sedge, and like the Spikerushes, it occupies large areas of the river floodplains between the grasslands and the paperbark swamps. Sedges also line the channels connecting the deeper water bodies and fringe the edges of permanent lagoons. In the sedgelands of the Top End, birds such as Comb-crested Jacanas (Irediparra gallinacea) and Brolgas (Grus rubicundus) nest.

Herbs

A number of small plants grow in the various habitats of the wetlands and like most other inhabitants of the region they are delicately tuned into the changes brought about by the arrival and departure of the monsoon. The Onion Lily (Crinum asiaticum) appears and flowers as flood plains begin to fill at the beginning of the wet season. It is common along the banks of water bodies, and also grows in the lowland forests.

The yellow flowered Ludwigia adscenndens, which has no common name, appears at the end of the wet season. It grows in the damp soil on the edge of the receding water, forming dense mats of vegetation. Around August it is usually attacked by large numbers of small black beetles and is soon completely devoured. Other species of plants also use the margins of flood plains as a habitat. One herb, Dentella dioceca, has minute leaves and small white flowers. It grows not only on the edges of swamps and lagoons, but on drying creek banks and channels.

Algae and other Small Organisms

Small organisms are at the base of the ecological pyramid, and most of the organisms on the wetlands depend, directly or indirectly, on their presence. Plankton is the term used to include all microscopic plants and animals that drift near the surface of marine and freshwater environments. Phytoplankton is the component that includes the green energy producers that are at the beginning of food chains. In freshwater it is made up mostly of single or multicelled green algae, once considered to be part of the plant kingdom, but now commonly included in a separate kingdom called Protista. Diatoms, which are algae enclosed in silica cases, are important single celled producers in both marine and freshwater habitats. The traditional term zooplankton refers to microorganisms that eat the algae, such as rotifers, and the carnivorous microorganisms that prey on the herbivores, such as microcrustaceans such as copepods and water fleas, Daphnia spp. It is all rather like a miniature jungle contained in a drop of water, but this invisible world is of immense importance to the whole wetlands system.

Birds in Wetlands

Waterbirds use many kinds of wetlands, including swamps, lagoons, mudflats, estuaries, bays and open beaches, freshwater and saltwater lakes, rivers, floodplain wetlands and dams.

Waterbirds include:

- Waterfowl such as ducks, geese and swans,
- Grebes,
- Pelicans,
- Cormorants,

- Ibises and spoonbills,
- Egrets and herons,
- Crakes, rails and waterhens,
- Terns and gulls,
- Shorebirds, which are also known as waders.

They fall into three main categories:

- Colonial nesting waterbirds require substantial floods to support large breeding events in floodplain wetlands. They include egrets, ibises, pelicans, cormorants and herons.
- Non-colonial waterbirds, including resident shorebird species, generally don't congregate to breed but are still dependent on wetlands for nesting and feeding habitat in which to raise their young. They include waterfowl, grebes, crakes, rails and waterhens.
- Migratory waterbirds, such as migratory shorebirds, use a range of wetlands to rest, feed and breed during their annual long journeys between wetlands in Australia and their breeding sites in the northern hemisphere.

Other native bird species also depend on wetlands for all or part of the life cycle. These include reed-warblers, grassbirds and birds of prey such as the swamp harrier and white-bellied sea eagle.

Royal spoonbill (Platalea regia), Macquarie Marshes Nature Reserve.

Birds that Live in Wetlands

Floodplain wetlands make up most of NSW's wetland area and provide important habitat for waterbirds. Many species depend on them for breeding. Some sites regularly support more than 20,000 waterbirds. This figure rises to more than 100,000 during large floods.

At those times the internationally significant Macquarie Marshes, Gwydir Wetlands and Narran Lakes support some of the largest breeding colonies of straw-necked ibis, intermediate egret and rufous night heron in southern Australia. Some waterbird species are thought to breed almost entirely within wetlands in the Murray–Darling Basin. They include glossy ibis, freckled duck and blue-billed duck.

Birds that Stop over in Wetlands

Many coastal and inland wetlands support a variety of migratory waterbird and shorebird species. These birds travel each year between their breeding grounds in the northern hemisphere and non-breeding habitats in Australia.

The Hunter Estuary Wetlands provide habitat for more than 100 species of waterbirds, including 45 species listed under international migratory waterbird agreements that Australia has with Japan, China and the Republic of Korea. The wetlands in Botany Bay support up to 34 migratory species, including the endangered little tern and an endangered shorebird community at Taren Point.

Myall Lakes can support up to 22 migratory species, including the eastern curlew, red-necked stint and bar-tailed godwit.

Need of Wetlands for Birds

There are three main reasons why birds need wetlands: for feeding, breeding and as a place to refuel and rest during migrations.

They don't necessarily stay in one wetland area, but will move between them. For example, many waterbirds move regularly to newly flooded habitats to feed and/or breed before that wetland dries out. Equally, semi-permanent, permanent and coastal wetlands provide refuge for birds when wetlands in other regions are dry for long periods.

Feeding and Roosting

In general, most waterbird species are dependent on aquatic food sources including wetland vegetation, aquatic invertebrates, fish and frogs.

For this reason they rely on different types of wetland:

- Fish-eaters include larger waterbirds such as pelicans, cormorants, terns generally depend on large open, deep waterbodies such as estuaries, lakes and waterholes.
- Plant-eaters include grazing waterfowl such as Australian wood ducks, black swans, magpie geese and plumed whistling-ducks. They're often found roosting on grassy banks of a wetland or 'bottom up' feeding on wetland plants.
- Invertebrate feeders include spoonbills, ibis, stilts, oystercatchers, dotterels and sandpipers. Mudflats and the shallow edges of waterbodies are rich feeding areas for these species.
- Diving ducks – such as musk dusks and hardheads – feed on both aquatic plants and invertebrates such as polychaetes (a type of worm), small molluscs and chironomids (midges or lake flies). They're attracted to wetlands with deep, open water.
- Shoreline foragers – such as bitterns, crakes, rails and swamphens – feed on invertebrates. They feed in reeds and rushes.
- Large waders, such as ibis, egrets and herons, feed on invertebrates, fish and amphibians, in shallow floodplain habitat including rushes, water couch and sedges.

Breeding

Heavy rainfall in inland NSW can cause widespread flooding and a large increase in waterbird numbers, especially in spring and summer. That's because flooding leads to higher levels of the food that waterbirds feed on. This in turn provides ideal breeding conditions for many waterbirds. Some species respond very quickly to heavy rainfall, moving large distances overnight to newly flooded habitats. For example, grey teals have been tracked with satellite transmitters moving 980 kilometres in 40 hours between rice-growing areas in southern NSW and Lake Eyre.

Migration

Migratory shorebirds rest and feed in NSW wetlands and other parts of Australia from September to March to prepare for their annual migrations to breeding grounds in the Arctic. These birds travel to breed in Alaska, Siberia, Mongolia, northern China or Japan, and then back again at the end of the breeding season. They fly continuously between stop-over sites, which need to be safe places to rest and refuel.

Migratory shorebirds such as sandpipers, dotterels and plovers also use inland wetlands when the habitat is suitable – for example as wetlands dry out after flooding to reveal shallow muddy feeding habitat.

Threats to Wetland Birds

There are 26 threatened waterbird species in NSW, including the Australian painted snipe, Australasian bittern, brolga, eastern curlew and freckled duck.

Many species are threatened by the loss and degradation of wetland habitats. This can be caused by:

- Clearing and draining wetlands for agricultural and urban development.
- Alterations to flooding patterns due to river regulation and climate change.
- Disturbance of feeding and roosting habitats by recreational users such as people fishing or walking, especially with dogs, or by feral animals such as foxes.

Other Animals in Wetlands

Fish that Live in Wetlands

Fish use wetlands for breeding, feeding and shelter, whether that's in coastal or inland environments.

Native Fish: Rivers and freshwater wetlands in NSW are home to 49 native fish species, ranging from carp gudgeons about 5 centimetres long to the iconic Murray cod, which can grow up to 1.8 metres in length and weigh up to 114 kilograms. Other species that regularly inhabit NSW freshwater wetlands include golden perch, Australian smelt, bony bream and Murray–Darling rainbowfish.

They can be found in many types of wetland, including lakes, floodplains, swamps, marshes, waterholes and billabongs.

Introduced Fish: Nine introduced or alien fish species are found in NSW freshwater rivers and wetlands. They include species such as the brown trout, rainbow trout and redfin perch, which were introduced to improve freshwater angling.

Others are now major threats to native fish and their habitats. These include the European carp, which was brought to Australia in the mid to late 1800s, and gambusia (also known as mosquito fish and plague minnow), which was introduced in the 1920s to control mosquitoes.

Marine Fish: Many species of marine fish rely on coastal wetlands to breed and nurture their young before returning to the open ocean.

These include most commercially and recreationally valuable species, such as flathead, mullet, bream, whiting, and shellfish such as prawns and crabs.

Need of Wetlands for Fish

1. Feeding and breeding: When an inland floodplain is inundated, many species of native fish colonise these nutrient-rich freshwater wetlands to feed and spawn. These areas then provide nursery habitats for juvenile fish.

   Coastal and marine wetlands are important breeding grounds and nursery areas for many kinds of saltwater fish. For example, areas of coastal saltmarsh are included on the list of Endangered Ecological Communities in New South Wales.

2. Moving from area to area: Fish use floodplains to scatter and move between permanent rivers or waterholes, the latter providing them with refuge during drought periods.

   With some species, like golden perch and silver perch, major spawning can occur in the river channel after flooding. This inundates adjoining floodplains, allowing juvenile fish to access wetland habitats.

3. Distribution of nutrients: Over-bank flows that inundate floodplain wetlands are also important for river-dwelling fish species such as the Murray cod and trout cod.

Rising floodwaters collect food and nutrients that have been deposited on the floodplain during drier periods and carry them back to the river as waters recede, and the fish can then feed on them.

Frogs that Live in Wetlands

Two-thirds of the frog species known in New South Wales depend on wetlands for their survival.

Alpine tree frog (Litoria verreauxii alpina).

Of the 71 frog species known in NSW, 47 are dependent on wetlands. Common wetland frogs include the striped marsh frog, brown-striped grass frog, spotted grass frog, green tree frog and red-eyed green tree frog. Because many frogs live in or around wetlands that are only intermittently wet, some have adapted to surviving long dry periods.

For example, 15 species of frog occur in the Macquarie Marshes and seven of those live underground during long dry periods. These include the striped burrowing frog, ornate burrowing frog, waterholding frog and crucifix toad.

Importance of Wetlands for Frogs

- Frogs and tadpoles have thin, porous skin, through which they absorb chemicals from the air and water. For this reason, more than any other terrestrial animal, they need water to survive. Most wetland frogs have very limited tolerances for drying.
- Frogs also need water to be able to reproduce. Local rainfall can create temporary pools suitable for some species to breed. However, most species depend on wetland flooding to breed, particularly in inland NSW, where water can be scarce for years.
- The majority of frogs in inland wetlands are most active in spring and summer and have a preference for wetlands with longer periods of water pooling. The timing of flooding also has a strong influence on which species are able to breed.
- Aquatic vegetation provides shelter for adult frogs and enables the growth of biofilms and organic matter, which are important food sources for tadpoles.

While most common frogs live in or around fresh water, many live in coastal waters, including the endangered green and golden bell frog. It thrives in water bodies that are unshaded, are free of predatory fish such as gambusia (plague minnow) and have grassy areas and sheltering sites nearby.

Mammals that Live in Wetlands

At least 9 species of mammal live in NSW wetlands, where they find plentiful supplies of their preferred foods.

There are at least nine species of mammal that rely wholly or partly on wetlands in New South Wales. They include the following native mammals:

- The swamp rat is found on the NSW coast and lives in heathlands, densely vegetated swamps in forests and in thick vegetation along rivers and streams.
- The common water rat is one of Australia's largest rodents and is usually found near permanent bodies of water. It lives in burrows in river and lake banks.
- The platypus lives in burrows in the banks of rivers, streams, creeks and ponds, ideally ones that are shaded by native vegetation.
- The fishing bat lives in tree hollows, caves, mines, culverts and under bridges, often close to water. It's also known as the southern myotis or large-footed myotis.

Other mammals that live around wetlands include the yellow-footed antechinus, common planigale, common blossom-bat, eastern chestnut mouse and pale field rat.

Need of Wetlands for Mammals

Mammals live in wetlands because they are adapted to the wet conditions and there is a plentiful supply of their preferred foods. For example:

- The swamp rat feeds on grasses, sedges, reeds, seeds and insects.
- The water rat feeds on a wide range of prey including large insects, crustaceans, mussels and fishes, and even frogs, lizards, small mammals and water birds.
- The platypus mainly feeds during the night on a wide variety of aquatic invertebrates, free-swimming organisms such as shrimps, swimming beetles, water bugs and tadpoles, and at times worms, freshwater pea mussels and snails.
- The fishing bat feeds on aquatic insects, small fish and flies close to the surface of rainforest streams or large lakes and reservoirs.

Reptiles in Wetlands

Many reptiles are dependent on NSW wetlands, including freshwater turtles, water skinks, snakes and water dragons. Some reptiles - such as freshwater turtles – live in wetlands for a large part of their life cycles. Others such as water skinks and red-bellied black snakes – rely on wetlands sometimes and spend the rest of their lives on dry land, usually quite close to streams or open water.

- Freshwater turtles: Freshwater turtles found in NSW wetlands include the eastern long-necked turtle, broad-shelled river turtle (found in rivers of Murray–Darling Basin) and the Murray turtle (found in rivers of the Murray–Darling Basin and associated drainages west of the Great Dividing Range).
- Water skinks: Five species of water skink have been found living in or close to NSW wetlands: the Blue Mountains water skink, the alpine water skink, the eastern water skink, the southern water skink and the swamp water skink.
- Snakes: Snakes that spend a lot of time around NSW wetlands and rivers include the red-bellied black snake, the Australian copperhead, the eastern tiger snake, the blue-bellied black snake and the rough-scaled snake.

- Eastern water dragon: The eastern water dragon lives in coastal waters, including marine wetlands. It is a good tree climber and likes to laze on branches overhanging the water.

Importance of Wetlands for Reptiles

Reptiles that use wetlands do so because they either live in water for much of their lives or rely on water for their survival. They also find plentiful food sources in wetlands, in the form of other animals.

Reptiles rely on wetlands in the following ways:

- Freshwater turtles use rivers, lakes and billabongs for feeding and breeding.
- Water skinks also rely on wetlands such as upland swamps for their food sources (insects, grubs and larvae) and as cover from predators. Some species, such as the alpine and Blue Mountains water skink, can survive at high altitudes – an unusual feat for cold-blooded animals.
- Snakes are attracted to rivers and wetlands because of the abundance of food in the form of frogs and waterbird eggs.
- The Eastern water dragon feeds on small reptiles, small mammals, frogs, insects, worms, molluscs, vegetation and fruit in or near the wetlands. It lays its eggs in a hole above the floodline of a wetland. It can also drop from branches into the water and hide from danger for up to 30 minutes.

Threats to Wetlands

Human activities cause wetland degradation and loss by changing water quality, quantity, and flow rates; increasing pollutant inputs; and changing species composition as a result of disturbance and the introduction of nonnative species.

Hydrologic Alterations

A wetland's characteristics evolve when hydrologic conditions cause the water table to saturate or inundate the soil for a certain amount of time each year. Any change in hydrology can significantly alter the soil chemistry and plant and animal communities. Common hydrologic alterations in wetland areas include:

- Deposition of fill material for development.
- Drainage for development, farming, and mosquito control.
- Dredging and stream channelization for navigation, development, and flood control.
- Diking and damming to form ponds and lakes.
- Diversion of flow to or from wetlands.
- Addition of impervious surfaces in the watershed, thereby increasing water and pollutant runoff into wetlands.

Pollution Inputs

Although wetlands are capable of absorbing pollutants from the surface water, there is a limit to their capacity to do so. The primary pollutants causing wet-l and degradation are sediment, fertilizer, human sewage, animal waste, road salts, pesticides, heavy metals, and selenium. Pollutants can originate from many sources, including:

- Runoff from urban, agricultural, silvicultural, and mining areas.
- Air pollution from cars, factories, and power plants.
- Old landfills and dumps that leak toxic substances.
- Marinas, where boats increase turbidity and release pollutants.

Vegetation Damage

Wetland plants are susceptible to degradation if subjected to hydrological changes and pollution inputs. Other activities that can impair wetland vegetation include:

- Grazing by domestic animals.
- Introduction of nonnative plants that compete with natives.
- Removal of vegetation for peat mining.

Conservation of Wetlands

Conservation of wetlands with the objective of protecting their biodiversity, specific biophysical characteristics and obtaining optimum benefits from them requires a major shift in polcies related land and water use. Joining the Ramsar Convention is only being a part of the international community to express solidarity with its objectives. Designation of a few wetlands under the Ramsar Convention and enlisting a few other large ones as important wetlands does not ensure the conservation of all wetlands. Studies show that many small wetlands together support more biological diversity than one large wetland though some species may occur only in large wetland. Just as drops make an ocean, each small wetland has a bit to contribute. No two wetlands are entirely similar like a photocopy or identical twins.

We need to address the real causes of wetland loss and degradation:

- First and foremost, both natural and human-made wetlands should be declared as specific land use category and their hydrological characteristics (sources and regimes) should be identified.
- Second, their conversion to any other land use or any reduction in their area or alteration in their water regime should be prohibited, except for strategic reasons after exploring other options and providing for compensatory measures.
- Third, the total biodiversity of all wetlands should be assessed and periodically monitored.
- Fourth, all ecosystem services of all wetlands should be assessed and valued in economic terms.
- All development projects such as those related to urban or industrial development, or those concerned with storage, diversion and abstraction of water from any source should consider all kinds of wetlands to be affected directly or indirectly, within the project area or far away from them. In the case of interventions on rivers, the hydrological changes may cascade down to the entire river course downstream (e.g. on floodplains). These projects should take into account especially the changes in biodiversity and the ecosystem services of wetlands, and their economic valuation should be integrated into the cost-benefit analysis of the project.
- The assessment of ecosystem services and their economic valuation should particularly address the benefits to the local community and their livelihoods. A few kilograms of rhizomes of lotus or the leafy shoots of Ipomoes aquatica may not be priced significantly but may have a high value for the local community in terms of vegetable use and nourishment a no cost to them.

References

- Wetland, science: britannica.com, Retrieved 5 February, 2019
- Ecology-of-wetland-ecosystems-water-substrate-and-17059765, library, knowledge, scitable: nature.com, Retrieved 19 April, 2019
- Birds, plants-and-animals-in-wetlands, wetlands, water: environment.nsw.gov.au, Retrieved 17 May, 2019
- Plants-and-animals-in-wetlands, wetlands, water: environment.nsw.gov.au, Retrieved 14 July, 2019

Permissions

All chapters in this book are published with permission under the Creative Commons Attribution Share Alike License or equivalent. Every chapter published in this book has been scrutinized by our experts. Their significance has been extensively debated. The topics covered herein carry significant information for a comprehensive understanding. They may even be implemented as practical applications or may be referred to as a beginning point for further studies.

We would like to thank the editorial team for lending their expertise to make the book truly unique. They have played a crucial role in the development of this book. Without their invaluable contributions this book wouldn't have been possible. They have made vital efforts to compile up to date information on the varied aspects of this subject to make this book a valuable addition to the collection of many professionals and students.

This book was conceptualized with the vision of imparting up-to-date and integrated information in this field. To ensure the same, a matchless editorial board was set up. Every individual on the board went through rigorous rounds of assessment to prove their worth. After which they invested a large part of their time researching and compiling the most relevant data for our readers.

The editorial board has been involved in producing this book since its inception. They have spent rigorous hours researching and exploring the diverse topics which have resulted in the successful publishing of this book. They have passed on their knowledge of decades through this book. To expedite this challenging task, the publisher supported the team at every step. A small team of assistant editors was also appointed to further simplify the editing procedure and attain best results for the readers.

Apart from the editorial board, the designing team has also invested a significant amount of their time in understanding the subject and creating the most relevant covers. They scrutinized every image to scout for the most suitable representation of the subject and create an appropriate cover for the book.

The publishing team has been an ardent support to the editorial, designing and production team. Their endless efforts to recruit the best for this project, has resulted in the accomplishment of this book. They are a veteran in the field of academics and their pool of knowledge is as vast as their experience in printing. Their expertise and guidance has proved useful at every step. Their uncompromising quality standards have made this book an exceptional effort. Their encouragement from time to time has been an inspiration for everyone.

The publisher and the editorial board hope that this book will prove to be a valuable piece of knowledge for students, practitioners and scholars across the globe.

Index

www.ingramcontent.com/pod-product-compliance
Lightning Source LLC
LaVergne TN
LVHW080655210926
848485LV00003B/381